SCIENCE AND TECHNOLOGY IN

20TH- CENTURY AMERICAN LIFE

Recent Titles in
The Greenwood Press "Daily Life Through History" Series

Civilians in Wartime Africa: From Slavery Days to the Rwandan Genocide
John Laband, editor

The Korean War
Paul M. Edwards

World War I
Jennifer D. Keene

Civilians in Wartime Early America: From the Colonial Era to the Civil War
David S. Heidler and Jeanne T. Heidler, editors

Civilians in Wartime Modern America: From the Indian Wars to the Vietnam War
David S. Heidler and Jeanne T. Heidler, editors

Civilians in Wartime Asia: From the Taiping Rebellion to the Vietnam War
Stewart Lone, editor

The French Revolution
James M. Anderson

Daily Life in Stuart England
Jeffrey Forgeng

The Revolutionary War
Charles P. Neimeyer

The American Army in Transition, 1865–1898
Michael L. Tate

Civilians in Wartime Europe, 1618–1900
Linda S. Frey and Marsha L. Frey, editors

The Vietnam War
James E. Westheider

SCIENCE AND TECHNOLOGY IN

20TH-CENTURY AMERICAN LIFE

CHRISTOPHER CUMO

The Greenwood Press "Daily Life Through History" Series

Science and Technology in Everyday Life

GREENWOOD PRESS
Westport, Connecticut • London

Library of Congress Cataloging-in-Publication Data

Cumo, Christopher.
    Science and technology in 20th-century American life /
Christopher Cumo.
        p. cm.—(Greenwood Press "Daily life through history"
series. Science and technology in everyday life, ISSN 1080–4749)
    Includes bibliographical references and index.
    ISBN: 978–0–313–33818–2 (alk. paper)
    1. Science—United States—History—20th century. 2. Technology—
United States—History—20th century. I. Title.
    Q127.U6C86 2007
    509.73′0904—dc22   2007023462

British Library Cataloguing in Publication Data is available.

Library of Congress Catalog Card Number: 2007023462
ISBN-13: 978–0–313–33818–2
ISSN: 1080–4749

First published in 2007

Greenwood Press, 88 Post Road West, Westport, CT 06881
An imprint of Greenwood Publishing Group, Inc.
www.greenwood.com

Printed in the United States of America

The paper used in this book complies with the
Permanent Paper Standard issued by the National
Information Standards Organization (Z39.48–1984).

10 9 8 7 6 5 4 3 2 1

CONTENTS

PREFACE

This book aims to describe how science and technology shaped the lives of ordinary Americans in the 20th century. Popular perception holds that science and technology are the two ends of a single pipeline. All one need do is invest in the science end of the pipeline to reap practical benefits from the technology end. In hopes of gaining funding, some scientists perpetuate this model of science and technology, but the reality is not so simple. Science and technology are not part of a single continuum but separate types of inquiry. Science differs from technology and from all other types of inquiry in its method. At the core of this method is the formation of a hypothesis that can be affirmed or falsified by a test. At times, the test of a hypothesis is an experiment. To test the hypothesis that organisms pass on to their offspring what we call genes, the Austrian monk Gregor Mendel conducted the experiment of crossbreeding varieties of peas. Yet not all science is experimental. By comparing different species of finches and other animals on the Galápagos Islands, British naturalist Charles Darwin arrived at the postulate that evolution by natural selection explains the current diversity of life. In this instance, observation and comparison took the place of experiment. Although science in all its manifestations shares the same method, not all science aims for the same end. Pure or theoretical science seeks knowledge of some component of reality for its own sake. Pure science does not seek a utilitarian end. The American physicist and Nobel laureate Albert Michelson attempted to measure the speed of light without the expectation that this knowledge would be useful to anyone. Occasionally, pure science has yielded a practical result even though this result was not the original aim. Albert Einstein arrived at the theoretical postulate that energy and matter are interconvertible. After 1939, however, physicists used this knowledge in constructing an atomic bomb. In contrast to pure science, applied science aims to derive a useful product. The applied scientist who crossbreeds varieties of wheat has the aim of high yield, disease resistance, or another agronomic trait. Technology, on the other hand, does not use the scientific method, because the generation of a hypothesis is not an essential feature of technology. It shares with applied science, however, the aim of deriving a useful product. In this context one should remember that technology is both process and product. The design and construction of an automobile is a process, whereas the car is a product. Both the process of manufacturing a car and the car itself are technology.

The aim of this book excludes pure science from consideration, not because it lacks merit but because its findings seldom trickle down to the masses. Einstein's principle of the interconvertibility of matter and energy is an exception. So is the premise of Darwin and his compatriot Alfred Russel Wallace that natural selection is the mechanism of evolution. From the outset, this claim sparked a controversy that continues to reverberate through the halls of our public schools. The decision to include or exclude evolution from the biology curriculum affects what students read, what questions they ask, and how they interpret evidence. These issues penetrate our everyday existence. To put the matter another way, evolution, whether as science or the antithesis of evangelical Christianity, has trickled down to all of us. In the main, however, pure science remains the province of intellectuals. To the extent that ordinary Americans take note of pure science at all, they tend to dismiss it as arcane—the work of those pedants the magazine *Ohio Farmer* derided as the "owl-eyed professors."

Applied science is another matter. Its origins in the United States can be traced to the heady early days of the republic when Thomas Jefferson, Benjamin Franklin, and George Washington united in praise of practical knowledge. They were conscious of creating a nation that trod a new path rather than the old dead ends of Europe. The intelligentsia of the Old World might while away the time on esoteric ideas, but Americans were too busy carving out a civilization in the New World to bother with abstruse hand-wringing. Science had to yield useful results to earn its keep in a land of farmers and shopkeepers. A leader of the American Philosophical Society, Franklin steered its resources toward the design of plows and had a hand in the founding of the Philadelphia Society for the Promotion of Useful Knowledge. In the spirit of Franklin, Jefferson designed his own plows and ran his estate at Monticello as an experimental farm. Like Jefferson, Washington conducted his own experiments, comparing the yields of different varieties of wheat on his farm. Implicit in this work was the conviction that any farmer could do his own experiments. Science was not a creed that Americans preached on Sunday only to ignore

Monday through Saturday. Rather, it was an activity that every American, regardless of class and education, could pursue. Science was more than an avocation. It was a virtue of a democratic society and an enterprise of the workaday world.

Only a short step separates the vision of science as an applied discipline from the belief that technology is an extension of science. Although this belief ignores the methodological distinction between science and technology, it serves to reinforce the utilitarian bent of both. The result in the United States is a blurring of the boundary between science and technology. The inventor Thomas Edison, the automaker Henry Ford, and the plant breeder Luther Burbank knew and respected one another. The fact that one might classify Edison and Ford as technicians and Burbank as an applied scientist matters less than the fact that all three sought to create useful products that American consumers would buy.

In short, Americans tend to define science as a utilitarian pursuit and to minimize the importance of theory. This narrowing of the bounds of science allows Americans to lump it in with technology. In the popular mind, *science* and *technology* are near synonyms for a creative process that yields practical results.

This book is mindful of the tendency of popular opinion to group together applied science and technology and of American capitalism to create mass markets for products. Accordingly, this book focuses on the products of applied science and technology that created a mass market and thereby shaped the lives of ordinary Americans in the 20th century: the automobile, dichlorodiphenyltrichloroethane (DDT), the microwave oven, the personal computer, and many other consumer goods. This book aims to recount how each of these products affected for good or ill the quality of American life in the 20th century. DDT, for example, benefited farmers by killing insects that otherwise would have eaten a portion of their crops. Consumers benefited as well, because the surplus of food made possible by DDT and other agrochemicals resulted in low prices. At the same time, DDT harmed the ecosystem by accumulating as a toxin in birds and mammals. This book does not praise or condemn DDT or the other products of applied science and technology but rather examines the multiplicity of their effects. As important as the products of applied science and technology are, this book also examines these endeavors as processes. DDT is not simply a product but a historical process that encompasses the manufacture of munitions in Europe, the testing of the insecticide by chemists and entomologists at the U.S. Department of Agriculture and the land-grant universities, the use of DDT by farmers throughout the United States, the opposition to DDT by environmentalists, and the ban on DDT by Congress. This book is therefore a history of how science and technology shaped the lives of ordinary Americans in the 20th century.

As a history, this book is a narrative in time and by topic. As a narrative in time, this book preserves the chronology of events as a way of organizing the past. As a topical narrative, this book groups chronology by topic. DDT, for example, functions as both topic and chronology of events, from the isolation of DDT as an insecticide in 1939 to its ban in the United States in 1972. Above this level of organization is the chapter. DDT is part of the chapter on agriculture; the automobile, of transportation; and the personal computer, of communication and media. Like its subtopics, each chapter organizes material by time and topic. The result, I hope, is a cohesive narrative of the main currents of science and technology and their effects on the lives of ordinary Americans in the 20th century.

INTRODUCTION

American science and technology originated not in Philadelphia or Boston but in Paris, where a band of intellectuals in the 18th century led the Enlightenment, a movement that upheld British physicist and mathematician Isaac Newton as the paragon of science. They saw in his unification of motion on earth and in the heavens by the law of gravity an affirmation of the grandeur of science. The philosophes of the Enlightenment, imbued with Newton's spirit of inquiry, believed that science was too important to be merely an activity of intellectuals. Science had the power to improve the lives of ordinary people. The philosophes believed that progress was the normal state of affairs and that science could be harnessed to better people's quality of life. Sure that science could achieve a utilitarian end, they embraced technology for its obvious practicality. Together, science and technology would transform the world, if not into a utopia, at least into a place where people were masters of their fate.

This faith in the progressive nature of science and technology sank roots in the nascent United States, whose leaders allied themselves with the philosophes. Thomas Jefferson and George Washington, heirs to the Enlightenment, envisioned the United States as a republic of yeomen and advocated that their countrymen use science and technology to improve agriculture. At first, science was a simple affair. Farmers grew different varieties of a crop to gauge their yield in an informal way and, with the same aim, used different mixes of fertilizers. Farmers also began observing the habits of insects and identified those that ate their crops. During the 19th century, the spread of agriculture through the Midwest and onto the Plains increased the demand for labor. One solution was to devise labor-saving technology. Cincinnati, Ohio, inventor Obed Hussey in 1833 and Rockbridge County, Virginia, inventor Cyrus McCormick in 1834 patented a horse-drawn reaper to cut grain, an improvement over the scythe. In 1837 Moline, Illinois, inventor John Deere unveiled a plow with a highly polished steel blade to cut through the tough prairie sod. Impressive as these achievements were, the conviction grew in the mid-19th century that organizations would make faster progress than the lone individual could. The advocates of science and technology on a grand scale won their first victory in 1862 when Congress created the U.S. Department of Agriculture (USDA) and the land-grant colleges. Congress intended these institutions to further science and technology for the benefit of the farmer. In keeping with this ideal, the USDA and land-grant colleges bred crops with the aim of high yield, experimented with crop rotations and fertilizers to find the best sequence of crops and to improve the fertility of soils, and in the late 19th century tested the first insecticides, which were compounds of lead and arsenic. To further scientific agriculture, Congress in 1887 passed the Hatch Act, giving each state $15,000 to establish and maintain an agricultural experiment station, the research arm of the land-grant colleges.

The march of agriculture westward intensified the need for a technology of transport to carry food to market and people across the land. After 1790, private companies built roads, an ancient technology. These roads and the wagons that traversed them were not an advancement over the roads of Mesoamerica and South America, Europe, Asia, and Africa. The canal was likewise an ancient technology, though one more efficient than roads: four horses could haul 1 1/2 tons eighteen miles a day by road but 100 tons twenty-four miles a day by canal. In the second quarter of the 19th century, several states dug canals to link the Hudson River with Lake Erie and Lake Erie with the Ohio River, making it possible to ship goods from New York City to New Orleans. The steam engine marked an advance over human and animal power. In the 18th century, British inventors Thomas Newcomen and James Watt built the first steam engines, which operated on the principles of thermal and kinetic energy. The engine boiled water, and the resulting steam drove a piston encased in a cylinder. In 1807 American inventor Robert Fulton harnessed the power of the steam engine, launching the first steamboat on the Hudson River. American engineers were quick to appreciate the potential of the steam engine to speed travel on land and, borrowing from Great Britain, built railroads in the United States after 1830. The mileage covered by railroads grew exponentially in the 19th century, and the locomotive came to symbolize the ability of technology to span distance.

The Enlightenment, fulfilling its promise in farming and transportation, was slow to affect medical science, where the beliefs of Greek physician Hippocrates persisted into the 19th century. He held that the body contained four humors: phlegm, blood, yellow bile, and black bile. The body that maintained these humors in balance was healthy. An imbalance afflicted a person with illness. Accordingly, a physician treated illness by seeking to restore balance—bleeding a person, for example, to reduce what was thought to be an excess of blood. Physicians identified vomiting and diarrhea as other means of restoring balance. In the 1820s American physician Hans Gram, a founder of homeopathic medicine, held that the body sought equilibrium without the intervention of a physician. Balance among the humors, he claimed, was the normal state of affairs and the body regulated the humors on its own. Consequently, Gram believed that rather than aggressively treating a patient, a physician should act in moderation, giving the body every opportunity to heal itself. He counseled physicians to prescribe medicine in small increments, usually 1/30th the standard dose as part of a minimalist approach to treating illness. Whereas Gram sought to modify the tenets of Hippocrates, American physician John Griscom preferred to jettison them for the medieval notion that a corrupt atmosphere causes disease. In 1844 Griscom published "Sanitary Conditions of the Laboring Population of New York," asserting that filth corrupted the atmosphere and so caused disease. He wanted local and state government to create public health boards to oversee the cleaning of water and streets, the disposal of garbage, and the building and maintenance of sewers. His ideas took hold in the 1860s after cholera swept U.S. cities and an interest in sanitary conditions was rekindled. New York City created a metropolitan board of health in 1866, as did Massachusetts in 1869. By then, American hospitals were revolutionizing surgical procedures. In 1846 American dentist William Morton used ether as an anesthetic, and after 1865 hospitals, following British physician Joseph Lister, sprayed surgeries with an antiseptic mist. The new germ theory of medicine underscored the importance of sanitary conditions and with it the need to sterilize surgical instruments. In 1862 French chemist Louis Pasteur determined that bacteria cause fermentation, a discovery that focused attention on microbes. In 1876 German bacteriologist Robert Koch, working with anthrax in cattle, demonstrated that pathogens cause disease. At the end of the century, medicine acquired an unprecedented ability to peer into the body. In 1895 German physicist Wilhelm Roentgen discovered that cathode rays produce a type of radiation that penetrates the body, exposing the underlying tissue and skeleton. He dubbed this radiation *X-ray,* and the next year American hospitals began using it.

American cities were quick to reap the benefits of the Enlightenment faith in the progress of science and technology. Electricity transformed the city in the 19th century, following from British physicist Michael Faraday's demonstration in 1833 that a magnet circling a piece of iron generates electricity. The revolving magnet was a simple generator that produced direct current and was the source of electricity for the first lights. In 1877 Cincinnati was the first city to use the electric light, when Nicolas Longworth installed an electric light in his home. The next year,

John Wanamaker's department store in Philadelphia was the first business to use electric lights. Wanamaker used them to attract shoppers in the evening, an improvement that brought nighttime shopping into vogue in the late 19th century. After 1880, cities began to illuminate their streets with electric lights. American inventor Thomas Edison quickened the spread of the technology, patenting in 1880 the incandescent lightbulb. By one estimate, the United States had some 25 million lightbulbs by the end of the 19th century. By then, alternating current, which could carry electricity long distances, was replacing direct current, and cities were building electric power plants to meet the urban demand for electricity. Electricity also facilitated urban transportation, and in 1888 Richmond, Virginia, became the first city to open an electric railway to take urbanites from home to work. By 1895 U.S. cities had roughly 850 electric railways, and in 1897 Boston opened the first subway. No longer dependent on the horse and buggy or on travel by foot or bicycle, the urbanite could live farther from work. The electric railway thereby stimulated the growth of suburbs. The technology of road building also improved transit in the city. By 1890 asphalt and brick roads were supplanting the uneven wood and granite street of the mid-19th century.

No less dramatic were the effects of technology on the home. By the mid-19th century, the spread of lumber mills gave the homebuilder uniform wooden planks with which to build frames, an improvement over the hand-hewn and variable planks of the early 19th century. Steel or cast-iron nails joined these planks, and the availability of glass allowed homebuilders to insert more and larger windows in a house. By 1890 the homebuilder dug the foundation of a house with a steam shovel, an innovation that quickened work, rather than a horse-drawn scraper. Linoleum rivaled wood for flooring, and stone, tile, or brick made up the walls. By the 1980s, roofs were covered in asphalt shingles or tile. The basement housed the furnace, a large stove connected to the registers by ducts, for central heating. Coal powered the furnace before the availability of natural gas. An alternative to the furnace was the steam-heated radiator. The kerosene lamp illuminated the home until the invention and spread of the electric light. In the mid-19th century, Americans did laundry by scrub board or a hand-cranked washer and wringer. By the 1890s the Laundromat, with its electric washers, gave women an alternative to the drudgery of washing clothes by hand. The flatiron, heated on the stove, removed wrinkles from clothes. Though housekeeping remained laborious, the homemaker could, by 1900, rely on technology to ease her burden.

As it had in the city and home, electricity revolutionized communication. Before the 19th century, information traveled only as fast as transportation could carry it. Two technologies, both powered by electricity, made possible instantaneous communication. The first, the telegraph, had its origins in the work of American inventor Samuel Morse. In 1837 he devised a system that transmitted electricity in a series of dots and dashes that was code for the letters of the alphabet. Morse patented the telegraph in 1840, and in 1844 he sent the first telegram from Washington, D.C., to Baltimore, Maryland. In 1858 the first transatlantic cable made it possible to send a telegram from Washington, D.C., to London. The second electric-powered communication technology, the telephone, made possible instantaneous transmission of the human voice. American inventor Alexander Graham Bell patented the telephone in 1876 and the next year incorporated the Bell Telephone Company. Bell operations director Theodore Vail grasped the telephone's potential to obliterate distance and sought to extend telephone lines as far as engineers could string line, in 1887 opening lines between New York City and Albany and New York City and Boston. Two other technologies quickened the transmission of the printed word. In 1829 American inventor William Burt patented the first typewriter. In 1867 American printer Christopher Sholes, lawyer Carlos Glidden, and engineer Samuel Soule introduced a commercial model and the next year patented the Qwerty keyboard. In 1873 Remington Arms Company, eager to expand its operations beyond the production of firearms, bought the patent of Sholes, Glidden, and Soule and in 1876 began manufacturing typewriters for a mass market. In 1885 German American inventor Ottmar Mergenthaler patented the linotype, an improvement over the printing press that quickened the setting of type.

Remington and other armaments manufacturers also felt the effects of technology. The musket, the principle firearm until the late 19th century, had been made more lethal by the introduction of the bayonet in the 17th century. But set against this lethality was the slowness of the matchlock, a type of musket. Around 1700, the armies of Europe and the militia of the American colonies replaced it with the flintlock, which a soldier fired by driving a piece of flint against a steel plate to ignite the powder, rather than having to wait for a strand of rope to burn down to the chamber, as was necessary with the matchlock. Around the same time, the introduction of the ramrod enabled a soldier to pack a charge into the barrel, and the cartridge eased the loading of a musket by giving a soldier both ball and powder in a single package. By the mid-18th century, improvements in metallurgy halved the weight of artillery, making it mobile, and the use of a backsight and a screw-driven wedge to raise and lower the elevation of the barrel increased the accuracy of cannons. In 1814 Robert Fulton armed a steamboat with cannons, and in 1843 the U.S. Navy launched the first propeller-driven warship. In 1862 Union and Confederate forces employed the first armored warships. The percussion cap, which came into use in the mid-19th century, allowed a soldier to fire the musket more rapidly than had been possible with the flintlock. Contemporaneously, soldiers began to use the rifle. By imparting a spin to a ball, the rifled barrel increased the accuracy of a shot over what had been possible with the smoothbore musket. After 1885, the magazine increased the rate of fire over that of the cartridge. Rather than pause after each shot to load another cartridge, a soldier simply opened the chamber of his rifle to push forward another round from the magazine. In 1884 American inventor Hiram Maxim unveiled the first automatic machine gun and French chemist Paul Vieille invented smokeless powder. After 1890, the U.S. Army replaced black powder with smokelesss powder and began using cannons with a recoil mechanism. By holding in place after being fired, the new cannon could be used repeatedly without the need for realignment. In the 1890s trinitrotoluene (TNT) replaced picric acid as an explosive. By 1900 technology had made armies capable of carnage on a vast scale.

The Enlightenment reached a crossroads in the work of Charles Darwin. The prophets of progress saw in natural selection a mechanism to ensure the continual improvement of the human species by sweeping away harmful traits and magnifying advantageous ones. Others perceived that evolution ultimately consigned a species to extinction, and so despaired of progress. It was natural for evolution, an idea rich in implications, to spark controversy. In the United States controversy arose at Harvard University, where botanist Asa Gray championed Darwin against zoologist Louis Agassiz's conviction that species were fixed entities and so do not evolve. At stake was the future of education, not merely in the ivory tower of academebut in the public schools. On the basis of a tenuous connection to the ideas of Darwin, his cousin, Francis Galton began in the late 1860s to wonder whether humans might guide their evolution by selective breeding. By increasing the procreation of the intelligent and industrious and reducing that of the stupid, indolent, and immoral, humans might improve the fitness of their species. The attempt to improve our species by regulating breeding, eugenics, affected the reproductive choices of women, particularly the poor and immigrants.

In 1900 American physicist Albert Michelson, looking back over a century of advancement, congratulated scientists at the annual meeting of the American Association for the Advancement of Science on the progress they had made. Michelson had in mind theoretical science, but the same was true of applied science and technology, which had penetrated the lives of ordinary Americans more deeply than at any time in the past. The food people ate, the way they traveled, and the homes in which they lived were the products of science and technology. The Enlightenment had bestowed its blessings on the United States by harnessing science and technology to yield practical results. Americans anticipated that they would reap even greater benefits in the 20th century.

CHRONOLOGY

1900   Walter Reed confirmed the mosquito *Aedes aegypti* as the carrier of yellow fever virus, the cause of yellow fever.

Three botanists independently rediscovered the three laws of heredity described in Gregor Mendel's paper on pea hybridization.

Plague struck Hawaii and San Francisco, California.

1901   Charles Hart and Charles Parr invented the gasoline tractor.

1902   Charles Wardell Stiles identified the worm that causes hookworm disease.

The first case of pellagra in the United States is reported.

Willis Carrier designed the first air conditioner.

1903   Wilbur and Orville Wright piloted the first airplane.

1904   New York City opened its subway to the public.

1905   Yellow fever killed 988 people in Louisiana.

The first gas station opened.

1906   Reginald Fessenden was the first to transmit the human voice by radio waves.

1907   James Murray Spangler invented the vacuum sweeper.

         Alan Campbell-Swinton and Boris Rosing independently invented television.

         Indiana became the first state to pass a law mandating the sterilization of criminals and the mentally deficient.

         Leo Baekeland invented plastic

1908   Michigan poured the first concrete road.

         Henry Ford began to manufacture the Model T.

         Chicago, Illinois, became the first city to chlorinate its sewage.

1909   Charles Davenport established the Eugenics Record Office.

         Alva J. Fisher, an employee of the Hurley Machine Company of Chicago, invented the Thor washing machine

         George Harrison Shull announced that the segregation of corn into inbred lines and the crossing of these lines would yield corn with hybrid vigor.

1910   Congress passed the Insecticide Act, requiring truth in labeling.

         Paul Ehrlich isolated the drug Salvarsan.

         Georges Claude invented the neon light.

1911   Detroit, Michigan, erected the first stop sign.

         General Electric manufactured the first refrigerator.

1913   Irving Langmuir invented a lightbulb with a tungsten filament.

         Henry Ford introduced the assembly line at his Highland Park factory.

1914   Elmer V. McCollum discovered vitamin A, the first vitamin to be discovered.

1915   Henry Ford introduced the Tracford, his first tractor.

1917   Donald F. Jones bred the first commercial hybrid corn.

1918   American soldiers faced their first gas attack in a battle against the Germans.

1919   Edwin George invented a gasoline lawn mower.

1920   Radio station KDKA broadcast the results of the presidential election.

1922   Houston began to operate the first automatic traffic light.

1923	Rochester, New York, became the first city to add iodine to drinking water.
1925	The Scopes Monkey Trial upheld a law against teaching evolution in the public schools of Tennessee.
1928	Alexander Fleming discovered the antibiotic penicillin.
	Henry Ford began to manufacture the Tri-Motor airplane.
1933	Boeing introduced the 247 airplane.
	President Franklin Roosevelt broadcast his first Fireside Chat by radio.
1934	Lockheed introduced the Model 10 Electra airplane.
	Transcontinental and Western Air made cinema standard on its flights.
1936	Donald Douglas introduced the DC-3 airplane.
1939	Swiss company J. R. Geigy isolated the insecticide DDT.
1943	Farmers in the Corn Belt planted hybrids on 90 percent of their acreage.
	Selman Waksman discovered the antibiotic streptomycin.
1945	Grand Rapids, Michigan; Newburgh, New York; and Evanston, Illinois became the first cities to add fluorine to drinking water.
	The United States used the atomic bomb against Japan.
1947	Bendix Corporation marketed the first washing machine with a spin cycle.
	Raytheon invented the microwave oven.
	Edwin Land invented the Polaroid camera.
1954	Hobie Alter made a surfboard from balsa wood, the first of its kind.
1955	Jonas Salk's polio vaccine became available for public use.
	The U.S. Navy launched the first nuclear-powered submarine, the Nautilus.
1956	Congress passed the Federal-Aid Highway Act
1957	The Soviet Union launched Sputnik, provoking fears in the United States that the Soviets had surpassed the Americans in science and technology.
1961	Steve Russell created the first video game, Spacewar.

1962	Rachel Carson published *Silent Spring.*
1963	Schwinn Bicycle Company manufactured the Sting-Ray, the prototype of the BMX bicycle.
1964	The U.S. surgeon general issued the first report that warned against cigarette smoking.
	Gordon and Smith manufactured a skateboard with a fiberglass board, the first of its kind.
1965	Vita Pakt Juice Company manufactured a skateboard with baked clay wheels, the first of its kind.
1968	Kenneth Cooper published *Aerobics.*
1969	Boeing began to manufacture the 747 airplane.
1970	Linus Pauling announced that megadoses of vitamin C might prevent the common cold.
1970–71	The Southern Corn Leaf Blight destroyed 15 percent of the U.S. corn crop.
1971	Marcian "Ted" Hoff invented the microchip.
1972	Congress banned the use of DDT in the United States.
	Mildred Cooper and Kenneth Cooper published *Aerobics for Women.*
1973	Creative Urethanes manufactured a skateboard with polyurethane wheels, the first of its kind.
1975	William Gates and Paul Allen founded Microsoft.
1976	Steven Jobs and Stephen Wozniak founded Apple.
1977	The U.S. Army began training its troops on video games.
1981	Specialized Bicycle manufactured the Stumpjumper, the first mountain bike.
1990	Tim Berners-Lee devised the World Wide Web.
1993	The U.S. Senate held hearings on the violence of video games.
1994	Id Software created the video game Doom.
1997	Monsanto marketed Bt corn, the first genetically engineered variety of corn.
1999	For the second time in the decade, the U.S. Senate held hearings on the violence of video games.

1

AGRICULTURE

In the 20th-century science and technology shaped the lives of farmers to a greater extent than at any time in the past. Technology at last replaced the horse as a source of traction on the farm. After 1901, farmers adopted the gasoline tractor and, as the decades passed, a series of motorized vehicles for planting, cultivating, and harvesting crops. Science transformed the relationship between farmer and plant. The new varieties of hybrid corn swept the Corn Belt in the 1930s and changed the way farmers used seed and heightened their dependence on science. The science of chemistry gave the farmer new insecticides and fertilizers and with them unprecedented control over nature as well as the responsibility to match these chemicals to the needs of the farm rather than to use them indiscriminately. More than a tender of plants, the farmer became in the 20th century an operator of machines, an applied chemist, and a field naturalist.

TRACTORS

The Era Before Tractors

In 1901 Secretary of Agriculture James "Tama" Wilson dismissed Thomas Edison's claim that technology would soon replace the horse on the farm. To be sure, Wilson acknowledged a role for new modes of transit. The electric railroad, Wilson believed, would entice urbanites to settle the countryside and ease the commute from farm to city. But Wilson doubted the electric railroad or other contraptions would dethrone the horse, which had provided traction on the farm for centuries. Wilson was quick to identify the flaws of the automobile, which could not, he judged, traverse uneven ground or drive through mud. Agriculture in the early 20th century seemed to demand more horses rather than fewer, particularly in the West, where farms were large and labor scarce. Farmers on the large tracts in the Dakotas, Washington, and Oregon used combines so heavy that they required as many as thirty-three teams of horses to pull them. These combines were themselves state-of-the-art technology, cutting grain in a swath of fourteen to twenty feet and threshing and bundling it. The demands of planting, cultivating, and harvesting crops led farmers either to maintain a stable of horses or to hire the job. The combine cut the stalk near the ground. Once cut, the grain entered the thresher, which shook seed from the stalk. The thresher channeled the stalk to a straw stack and the grain to sacks. Teams of harvesters began work in June, moving north with the harvest to the Dakotas and to Montana by September. The combine seemed to prove that the horse and technology fed off each other and formed an interlocking system.

Integral to technology at the turn of the century, the horse shaped labor, the use of farmland, and the working lives of farmers. The horse could not be switched on and off like a lightbulb but required year-round care. The family short of labor hired a man to organize horses for work and to care for them. The horse thus heightened the demand for labor and kept men on the farm who might otherwise have sought work in the cities. The horse also commandeered its share of the harvest. The farmer had to set aside land to feed his horses: in parts of the Midwest he grew oats for his horses on as much as one-third of the land. What the farmer did not feed his own horses he sold as fodder to farmers from the South and West. Under these circumstances, one would have been rash to join Edison in forecasting the disappearance of the horse.

The Steam Tractor and Automobile

Early efforts to supplant the horse were not promising. By 1900 dozens of small manufacturers churned out some 5,000 steam tractors a year. Large and heavy, the steam tractor carried a quantity of water and coal. A gang of four men guided the tractor, stoked the engine with coal, and replenished the water. Although the steam tractor hitched to a combine was capable of harvesting 100 acres a day, the tractor was difficult to maneuver and had a large turning radius. Only with difficulty could the steam tractor cover uneven ground, and it bogged down in the mud. Ravenous in its consumption of labor as well as of coal and water, the steam tractor maintained the large workforce that had been a feature of agriculture in the era of the horse.

No better than the steam engine in its first manifestation on the farm was the gasoline engine. Manufacturers in the first years of the 20th century filled the pages of *Implement Trade Journal*, a weekly whose subscribers

In addition to automobiles and airplanes, Henry Ford manufactured tractors. His signature model was the Fordson tractor. Courtesy: Library of Congress.

were largely farmers, with ads for stationary gasoline engines, but these could not be a source of traction. Mounted on a chassis, the gasoline engine instead took to the road as the automobile. As early as 1903, the year Henry Ford founded the Ford Motor Company, farmers hitched their plows to the automobile. They could not, however, walk fast enough to keep pace with it, and the plow skimmed the ground rather than digging a furrow.

The Gasoline Tractor

Eschewing the automobile, Iowa inventors Charles Hart and Charles Parr in 1901 built the first gasoline tractor with the wide frame of the steam tractor, a low center of gravity, and low gearing. Like the steam tractor, the Hart-Parr tractor was heavy at 10 tons and had a wide turning radius. By 1906 Hart and Parr had pared the weight to 5 tons, increased horsepower to thirty, and began using the term *tractor* for their machine. As did Ford in the auto industry, Hart and Parr aimed to mass-produce the tractor and to standardize parts. By one estimate, more than thirty firms were manufacturing some 2,000 tractors a year by 1909. The first tractors, with studded steel wheels and no rubber tires, sacrificed comfort for traction and did not immediately replace the horse. Rather than decrease, the number of horses on U.S. farms increased from roughly 4 million in 1910 to 4.5 million in 1916.

Two events turned farmers toward the tractor. First, epidemics of spinal meningitis and blackleg swept American stables in the late 1910s, leaving farmers short of their traditional source of traction. Second, World War I heightened the demand for labor and food. The Allies, desperate for horses to pull materiel and supplies, inundated the United States with orders. The Avery Company of Peoria, Illinois, was one company among many to turn the war to its advantage, in 1914 declaring the sale of horses to the Allies a patriotic duty—a shrewd act given the high price of both horses and feed. The proceeds of these sales, the Avery Company hoped, would go to buy its tractors. The war siphoned off not only horses but also men, who entered the military or took jobs in the city at greater rates. Short of horses and laborers, American farms went from having 4,000 tractors in 1911 to 250,000 in 1920.

By then, John Deere, International Harvester, and Ford were vying for the market. Henry Ford, who had grown up on a farm and sought to improve agriculture as he had travel, in 1915 built his first tractor, the Tracford, boasting that it could do the work of twelve horses and touting its simplicity, durability, and ease of operation. Ford priced it at $250, nearly half the $495 that Happy Farm Tractor Company, for example, charged and less than the $325 of the Model T. Ford unveiled the Fordson tractor in 1917, selling more than 7,000 tractors that year and more than any other manufacturer in 1918. International Harvester rather than Ford, however, made the real innovations, building in the 1920s the Farmall, a narrow tractor that had a small turning radius and so could cultivate and harvest the row crops of the Midwest and South, not merely the wheat of the West. In the 1920s the corn picker complemented the tractor in the Midwest and mechanized the last phase of corn culture.

The tractor came with an ideology of efficiency that shaped the lives of farmers. The agricultural colleges and experiment stations had in the 19th century urged the farmer to be scientific in his work, by which they meant a willingness to apply the results of experiments on the farm. Now, in the era of the tractor, the colleges and experiment stations wanted farmers to be businessmen in addition to receptors of science. Whereas the horse could advance both work and leisure, the tractor could further only the industrialization of agriculture. The tractor enabled farmers to cultivate more land than had been possible with the horse, and with an increase in the scale of operations came the need to compartmentalize every component of work. Gone was the woman of the prairie who broke the sod alongside her husband. He now devolved on her the record keeping necessary to analyze every facet of the business and to satisfy creditors who had loaned them the money to buy a tractor and its associated gadgets. Her work was in the house and his on the land. Gone, too, were the days when children had to help with farm work, particularly during planting and harvest, tasks the tractor could manage on its own. Instead, children spent more time in school, and some even went to college. The towns and cities that had drawn men and women to the factory during World War I now enticed the young to their college campuses. The agricultural colleges were both a source of information for farmers and a lure to their children. With a mixture of pride and regret, parents watched their children graduate from these colleges as engineers and entomologists rather than return to the land as farmers.

With the industrialization of agriculture, the tractor shaped labor to its rhythm. The horse had walked no more than two miles an hour, the limit at which a human can walk without undue fatigue, but the tractor covered ground at an unremitting five miles an hour without the need to rest. The tractor kept the farmer on the move for days that stretched to eighteen hours. The routine of planting or another chore came to resemble the monotony of the big rig on an unvarying stretch of highway. Farmers suffered back pain from jostling, particularly before the introduction of rubber tires on the tractor, and hearing loss from the drone of the engine. Bereft of the romanticism of the horse, the tractor made the farmer more an operator of machines than a tender of plants. Business acumen and an attention to commodity prices meant more in the machine age than an intuitive ability to gauge the time at which a corn plant silks.

The tractor sharpened the distinction between owner and worker. The freeholder of the Midwest and West owned a tractor, whereas the tenant and sharecropper of the South, too poor to get credit from a bank, made do with the horse or the mule, a hybrid of the horse and donkey. The migrant laborer likewise was too poor to own a tractor. On the bottom rungs of the agricultural ladder, the tenant, sharecropper, and landless laborer failed to reap the rewards of the machine age.

The pace of mechanization slackened during the Great Depression, with the number of tractors in the United States holding constant at roughly 1 million between 1931 and 1934. Their incomes dwindling, farmers had little to invest in technology during these years. Between 1930 and 1935, investment in farm equipment fell by $336 million as labor became the substitute for technology and capital. Men who had lost their job in a factory returned to relatives in the countryside to work part-time. Awash in labor, farmers had neither the money nor an immediate incentive to adopt new technology.

The production of materiel during World War II lifted the United States out of the Great Depression and restored farmers to prosperity. Those who had forgone technology during the 1930s out of necessity were able to buy the latest implements: tractors with rubber tires and the ability to cover five miles an hour in the field and twenty on the road, disk plows, eight-row planters, and corn and cotton pickers. In the 1960s agricultural engineers developed machines that plowed, planted, and applied fertilizer in a single pass. In the 1970s tractors increased from sixty to two hundred horsepower. In the 1980s air-conditioned cabs became standard, and in the 1990s the first use of satellites enabled farmers to plow, plant, cultivate, and harvest in perfectly straight rows that do not overlap. Computers told farmers the yield of a crop while they were harvesting it and projected the amount and type of fertilizer, insecticide, and herbicide they would need the next year.

These technologies complicated the lives of farmers. The agricultural colleges and experiment stations urged them to modernize by adopting the newest tractor or combine, but these purchases indebted farmers, increasing their anxiety about their ability to repay loans in an era of low commodity prices. This anxiety heightened for farmers who could not insulate themselves from global forces. American farmers in the late 20th century could do nothing to influence the price of wheat in Argentina, and yet a bumper harvest there might diminish the price of wheat in North Dakota. Indebted and buffeted by forces outside their control, farmers were also beset by technical challenges. They needed specialized knowledge to understand the increasingly technical reports of agribusiness firms and to use the latest software. The lives of farmers in 2000 focused on reading technical literature, troubleshooting software, keeping abreast of commodity prices and projections for the future, and adhering to the maintenance schedule for a twenty-four-row planter. In the 20th century, technology had made the farmer a manager of information and implements.

HYBRID CORN

The Era Before Hybrid Corn

Since the 19th century, the U.S. Department of Agriculture, the agricultural colleges, and the experiment stations had bred crops for yield and resistance to diseases, insects, and drought, but by 1900 this work had not affected the lives of most farmers. They judged a crop by inspection, saving seed for planting the next year from plants that yielded well or manifested no damage from insects or disease. The farmer saved several ears of corn, for example, through the winter. A few weeks before planting he wrapped a single kernel from each ear in a wet cloth, planting in the spring seed from the ears whose kernels germinated and discarding those ears whose kernels did not germinate. This brand of agriculture required no science, only a keen eye and attention to detail.

Much of the efforts of scientists and farmers concentrated on corn, an indigenous grass that humans had grown for millennia. In the early 20th century, the Corn Show swept the Corn Belt of the Midwest. An annual contest analogous to the county fair, the Corn Show trumpeted the virtues of rural life and clean living and was as much about civic boosterism as it was about judging corn. The Corn Show substituted beauty for economics, eschewing yield as the criterion for evaluating corn. Rather, judges prized large ears with straight rows, full and uniform kernels, and no bird or insect damage. The winner claimed bragging rights in the county.

Agricultural scientists made the rounds at the Corn Show, renewing acquaintances and sharing their latest research in an informal setting. Some scientists championed the Corn Show for making farmers careful observers of their crop, but others derided the emphasis on aesthetics. In their view yield was the only measure of corn that mattered, and they set to work with the aim of breeding high-yielding varieties regardless of how these varieties looked.

In 1905 agronomist Cyril George Hopkins at the Illinois Agricultural Experiment Station devised the ear-to-row method of breeding corn. Hopkins collected ears from several corn plants, planting the seed from each in its own row. Because all plants in a row were siblings Hopkins could easily maintain the pedigree of each plant. Hopkins hoped that by keeping seed from only the highest-yielding row and repeating the ear-to-row method he would select for corn with progressively higher yield. Although the ear-to-row method allowed Hopkins to derive corn with high protein and oil, he made no headway in yield.

The Advent of Genetics and the Science of Hybrid Corn

Other scientists, impatient with the ear-to-row method, gravitated to the new science of genetics. In 1900 three European botanists independently rediscovered the paper Austrian monk Gregor Mendel had published in 1866 on pea hybridization. Mendel had shown that particles (genes) code for traits in pea plants, and after 1900 scientists extended his insight to other plants and animals. American geneticist George Harrison Shull, in Cold Spring Harbor, New York, chose to work with corn because it differed from peas in that it cross-fertilized rather than self-fertilized, and Shull wished to know whether Mendel's laws of inheritance held for cross-fertilizing as well as inbreeding populations. The crucial insight came in 1909 when Shull realized that inbreeding would reduce corn to several uniform populations, with each population similar to a variety of peas in its uniformity of genes. A breeder could then hybridize corn as Mendel had hybridized peas.

Hybridization held out the possibility of imparting heterosis, or hybrid vigor, to corn plants. For centuries humans had known that hybrids—the mule, for example—have greater vigor than their parents. In corn, hybrid vigor might display itself as high yield, resistance to diseases, insects and drought, stalk strength, or some combination of desirable traits.

Although simple in conception, the practice of breeding corn for hybrid vigor challenged a generation of agronomists. The problem lies in the biology of corn, which, as we have seen, cross-fertilizes rather than self-fertilizes. The tassel, which contains the pollen, and the silk, which holds the ovule, are far apart in corn, and wind carries the pollen from one plant to the silk of another. To inbreed corn, an agronomist or farmer or anyone who knows the anatomy of a corn plant must cover the tassel and silk of a plant to prevent pollen from one plant from blowing onto the silk of another. When the tassel is mature, the breeder collects the pollen and spreads it on the silk of the same plant. Several generations of inbreeding produce homozygous lines that breed true, as a variety of peas does. In turning the natural process of crossbreeding on its head, however, inbreeding weakens corn, producing scrawny ears with fewer seeds than the parents. This paucity gives the breeder enough seed to make a cross on a tiny plot but too little to make a cross on the scale that farmers need for their fields. As long as the seed yield was small, hybrid corn remained a curiosity rather than a commercial venture. But in 1917 agronomist Donald F. Jones at the Connecticut Agricultural Experiment Station got past the seed bottleneck by crossing four inbred lines over two generations. Jones crossed the four inbreds in two groups of two to yield two hybrids, and then crossed these two hybrids to produce one second-generation hybrid. Because this method required two generations of breeding, Jones called this hybrid a double cross to distinguish it from the single cross, a hybrid produced by one cross of two inbreds over one generation.

The Work of Making a Hybrid

The tedious job of making hybrid crosses fell to rural youth, shaping their lives during summer vacation from school. To make a hybrid cross, the breeder designated one row of corn male and its neighboring row female. To ensure that only plants in the male row distributed pollen, the breeder hired people to remove the tassel from plants in each female row. In the 1920s seed companies, first in Connecticut and then in Iowa, Illinois, and other states in the Corn Belt, began crossing corn on a large scale and needed a large labor force analogous to the landless laborers who were a prominent component of American agriculture. Like the work of picking fruit in the citrus groves in Florida and California, the detasseling of corn in Iowa was seasonal work that paid little, offered no benefits, and required few skills beyond stamina and a knowledge of the anatomy of a corn plant. The job of detasseling corn, poorly paid as it was, did not at first attract men and so fell to high school students, who were free summers when the job had to be done and whose income helped stretch the family budget. The task of detasseling corn, like the picking of fruit, had to be done rapidly in those few weeks when the tassels were emerging but before they began to disperse pollen. The need for rapidity kept crews of detasselers in the field for long, monotonous days of walking down the rows of plants designated female, checking each plant, and removing the tassel from all plants from which it had emerged. The job required repeated passes along each row during the critical weeks to ensure the removal of all tassels, particularly those that emerged late. The job, requiring a detasseler to be on his feet all day and to reach high on each plant to remove a tassel in a repetitive manner, tired the legs, back, and shoulders. In the hot weather common during summers in the Corn Belt, the job was especially onerous.

Seed companies began in the late 1920s to introduce gasoline-powered carts to carry workers down rows of corn. A detasseler stood atop a cart to do his job. Although this method reduced fatigue in the legs and lowered the height to which one had to reach to remove the tassel, carts quickened the pace of work, in the manner of the assembly line in the auto industry. The result was greater monotony, greater pressure, and less time to inspect each corn plant and to remove the tassel on plants from which it had emerged. The seed companies in effect created a type of Fordism in the fields. Also like the assembly line, the introduction of the cart in corn fields increased the turnover of workers and might have caused unrest but for the Great Depression. The men who lost factory jobs and returned to the countryside, desperate for work, competed with high school students for the job of detasseler. The seed companies, thanks to the Depression, had more labor than they needed and so had no incentive to improve working conditions and raise pay.

The Spread of Hybrid Corn

The seed companies were among the few to prosper during the Depression and in doing so revolutionized agriculture. Hybrids were more drought tolerant than traditional varieties, a fact the arid years 1934 and 1936 underscored. Farmers who planted traditional varieties in these years watched their corn wither and the hybrids of their neighbors survive. Hybrids had proven their worth and attracted farmers en masse. Whereas they planted hybrids on only 1 percent of acreage in the Corn Belt in 1933, farmers planted them on 90 percent of acreage in 1943.

Hybrids also altered the working lives of farmers. The farmer who had saved seed for planting the next year found himself at a loss with hybrids. If he saved seed from a hybrid he lost rather than gained yield because second-generation seed yields poorly. Discouraged by this result, farmers in the 1920s had abandoned hybrids, preferring to save seed for planting from traditional varieties. But the drought of 1934 and 1936, as previously mentioned, lured farmers to hybrids, this time for good. Unable to save seed from hybrids, the farmer now had to buy new seed each year. The idea of obsolescence in the auto industry reached its apex in the cornfield. As never before, the farmer found himself dependent on scientists and seed companies. Each year, the agricultural colleges and seed companies developed new hybrids, leaving the farmer at a loss as to which to plant. He turned to the county extension agent and representatives of seed companies for advice. The most enterprising farmers contracted with seed companies to grow inbreds for crossing on their farms, deepening further farmers' dependence on seed companies and science. The county extension agent worked with farmers in the field and held meetings in nearby towns to help them decide which hybrids to plant and how best to grow them. Hybrids made science more relevant to farmers than at any time in the past.

Moreover, hybrids reinforced the value of technology on the farm. The stalks of traditional varieties tended to bow under the weight of ears of corn. The stalks of hybrids, however, were strong and stayed straight, making hybrids easier than traditional varieties to harvest by machine. Hybrids thus hastened the spread of the mechanical harvester, introduced in the 1920s, throughout the Corn Belt.

In the 1950s seed companies began to replace the double cross with the single cross hybrid. The single cross simplified the breeding of corn by producing a commercial hybrid in only one generation. Scientists and seed companies were able to switch to the single cross hybrid because, after decades of research, they had inbreds that yielded enough seed to make a cross on the large scale necessary to produce hybrid seed for farmers throughout the United States.

The switch to single cross hybrids coincided with the discovery in the late 1940s of genes that made corn sterile and of other genes that restored fertility to corn. Although it may seem counterintuitive to produce a sterile variety of corn, its use revolutionized the breeding of hybrids. Whereas, before 1950 seed companies had hired adolescents to detassel corn, between 1950 and 1970 they dispensed with this labor by using the sterile inbred as the female line because the sterile line produced no pollen. By crossing the male sterile line with an inbred containing the genes that restored fertility, seed companies were able to produce a fertile single cross hybrid. The use of the male sterile line in hybrid crosses abruptly made obsolete the job of detasseler and thus deprived high school students in the Corn Belt of summer work and farm families of extra income.

In using sterile male lines, seed companies unwittingly courted disaster. The Texas cytoplasmic sterile line, the most widely used sterile male inbred in hybrid crosses, was vulnerable to a new fungal disease, the Southern Corn Leaf Blight, that arose in the Philippines in 1969 and swept the United States in 1970 and 1971. The blight cost farmers nationwide 15 percent of their corn crop, and some corn growers along the Mississippi and Ohio rivers lost their entire crop. Farmers turned to their extension agents for sympathy and to scientists for new varieties of corn. In breeding new varieties, the seed companies discarded the Texas line. Seed company Pioneer Hi-Bred eliminated the use of sterile male lines altogether and returned to hiring people to detassel corn. After an absence of twenty years, high school students once more worked as detasselers during summers.

The Bioengineering of Corn

The Southern Corn Leaf Blight heightened the emphasis on breeding disease-resistant corn. With this aim came the corollary of breeding

insect-resistant corn. Traditional breeding required the scientist to identify an inbred resistant to an insect or disease, then to cross it with another inbred to produce a resistant hybrid. The defect of this method is
the breeder's inability to transfer only those genes that confer resistance
from one plant to another. Rather, in making a cross, the breeder transfers
half the plant's genome. The result will be a hybrid with the advantageous
trait of resistance to an insect or disease but with other traits as well, some
benign and others disadvantageous: late maturation, poor germination,
little pollen, grain with low protein, and the like. Faced with these imper-
fections, the breeder must cross a variety of corn with other varieties over
several generations until he achieves the right combination of advantageous traits and minimizes disadvantageous ones. The process is tedious
and consumes years. In the 1970s, however, scientists learned to extract
sequences of nucleotide bases from one plant or organism and to insert
them into another. A period of experimentation followed, and in 1997
agrochemical company Monsanto began selling a variety of corn with
genes from the bacterium *Bacillus thuringiensis* that code for production of
a chemical toxic to the European Corn Borer, a pest of corn since the 1910s.
Bt corn allowed farmers to reduce the use of insecticides, saving them
money and minimizing the degradation of the environment. But in 1999
Cornell University reported that pollen from Bt corn kills monarch butterflies, igniting a debate over the safety of genetically engineered crops.
Stung by criticism, Monsanto withdrew Bt corn, leaving farmers little alternative to insecticides.

THE CHEMISTRY OF FARMING

The Chemistry of Insecticides and Fertilizers in the Early 20th Century

From an early date, chemistry shaped the lives of farmers in the United
States. By 1900 they used both insecticides and fertilizers. Insecticides in
these early years of the 20th century were compounds of lead and arsenic,
and although lethal to insects, they degraded quickly in the environment,
requiring farmers to apply an insecticide at frequent intervals throughout
the growing season. In an effort to improve soil fertility, farmers also applied both manure and fertilizers to their land. In an era when scientists
were eager to equate farming with business, the proponents of scientific
agriculture put a premium on efficiency and scoffed at the farmer who
wasted manure. Gone were the days when he let it accumulate in barns
or disposed of it in streams. Several companies manufactured manure
spreaders to ease the chore of fertilizing the soil, and agricultural peri-
odicals touted the use of manure on the farm. Manure was not by itself
the solution to the problem of soil exhaustion, scientists believed, but part
of a larger scheme that also included chemical fertilizers. Agricultural
chemists in the early 20th century focused on nitrogen, phosphorus, and

potassium, the holy trinity of nutrients for plant growth, and advocated the use of fertilizers with one or more of these elements. Farmers had a choice of innumerable brands, and, despite the entreaties of farmers, scientists were careful not to endorse one product over another, only to advocate truth in the labeling of contents. One scientist, typical in his opinion, recommended that farmers use a fertilizer, regardless of the brand, with nitrate for nitrogen, potash for potassium, and acid phosphate for phosphorus. To these chemicals scientists urged farmers to add calcium carbonate to reduce soil acidity.

In the first years of the 20th century, the application of insecticides and fertilizers required the labor of both farmer and horse. The farmer stored insecticides as liquid in tanks, loading them on a horse-drawn wagon for application on crops or fruit trees. Farmers worked in pairs to spray an insecticide: one pumped it from the tank and the other directed it as a mist through a nozzle. In contrast to the spraying of an insecticide, one farmer could fertilize the soil without the aid of a second. Also unlike insecticides, fertilizers before World War II were exclusively solid rather than liquid or gas. Farmers loaded sacks of fertilizer into a horse-drawn spreader, which broadcast it on a field in preparation for planting.

Alert to the danger of fraud, the farmer wanted scientists to regulate the chemical industry, but scientists believed these police duties a poor use of their time and so resisted the move toward regulation. The spirit of the Progressive Era, attuned to the potential of science to benefit the lives of Americans, sided with farmers, and among a stream of legislation Congress passed the Insecticide Act (1910), which required truth in the labeling of insecticides. State legislatures passed similar laws to cover the manufacture and labeling of fertilizers.

In 1910, the year Congress passed the Insecticide Act, manufacturers of insecticides in the United States grossed roughly $20 million. The leading sellers—Paris green, which was a dye, and lead arsenate—felt the effects of World War I. With acetic acid, an ingredient in Paris green, in shortage in 1917 and 1918, the supply of the insecticide fell below demand. Alert to the problem of supply, the U.S. Department of Agriculture urged farmers to switch from Paris green to lead arsenate. At that time, the demand created by World War I for shot, flares, and the poison diphenylchloroarsine, meant that the U.S. capacity to produce arsenic was increasing, and the production of lead arsenate was boosted in turn. Other arsenic compounds found a market as well. In 1917 USDA entomologist Bert Coad discovered that calcium arsenate was lethal against the boll weevil. The news catapulted the insecticide ahead of Paris green in sales, and by 1923 production of calcium arsenate exceeded 13 million pounds. Southern farmers, like their counterparts in the North, had entered the era of modern chemistry. To this list of insecticides USDA entomologist E. B. Blakeslee in 1919 added paradichlorobenzene, a byproduct of the manufacture of the explosive picric acid, and production of PDB rose from 131,000 pounds in 1919 to 402,000 in 1921 and to 14 million in 1940.

DDT

In 1939 Swiss chemical company J. R. Geigy made one of the epochal discoveries of the 20th century, isolating DDT and identifying it as an insecticide. DDT was a new kind of insecticide, killing on contact rather than being lethal to insects only upon ingestion. Moreover, DDT persisted in the environment weeks, even months, longer than the previous generation of insecticides, reducing the number of applications farmers needed to make. In 1942 Geigy gave a sample of DDT to the USDA, which in turn shared the insecticide with the agricultural colleges and experiment stations. Scientists sprayed DDT on barns, which farmers toured in astonishment at the absence of insects. Farmers spread DDT as a powder by airplane and as a liquid by mechanical sprayer hitched to a tractor. Amplifying the success of DDT in the 1940s was the new air-blast sprayer, which distributed a mist of pure DDT rather than a mixture of water and insecticide, as the previous generation of sprayers had.

In their enthusiasm for DDT, farmers abandoned other means of controlling insects. Farmers who had rotated corn with other crops to reduce the population of the western corn rootworm heeded the advice of the Nebraska Agricultural Experiment Station in 1948 to replace rotation with DDT. The growers of wheat followed suit, transforming the West and Midwest into areas of wheat and corn monoculture. Farmers likewise abandoned the plowing under of crop residue in the fall, a practice that reduced the plant debris in which insect eggs could pass the winter. Alert to the potential of chemicals similar to DDT in structure and effect, chemists discovered or synthesized more than twenty new insecticides between 1945 and 1953, and sales of insecticides in the United States rose from $9.2 million in 1939 to $174.6 million in 1954.

Apple growers demonstrated the magnitude of savings DDT made possible. Before the use of DDT, the codling moth threatened, in the estimation of one entomologist, to drive apple growers out of business. Part of the problem was the cost of insecticides and the labor to apply them, both of which had risen in Washington from $12.98 per acre in 1910 to $60.92 per acre in 1940. Requiring fewer applications than the old insecticides, DDT dropped these costs to $53.61 per acre in 1950. Whereas the old insecticides had been tallied at 45 percent of the cost of growing apples in 1940, DDT diminished the cost to 10 percent in 1950.

Yet DDT was no panacea. As early as 1944, Washington entomologists E. J. Newcomer and F. P. Dean observed that DDT, by killing insects indiscriminately, killed the predators of mites, allowing the population of mites to swell in apple orchards. Further work revealed the same phenomenon for the wooly apple aphid. If DDT alone was insufficient to kill all insects and arachnids, farmers and entomologists added other insecticides to their arsenal. Apple growers began in the 1940s to spray parathion to kill mites and TEPP and BHC to kill aphids in addition to DDT to kill moths. Farmers who had in 1944 gotten by with few applications of DDT found themselves by the end of the decade on a treadmill, spraying DDT and its kin more often—twelve to twenty times a year—to kill an assortment of insects and mites. Farmers purchased the desideratum of an insect-free farm at an increasingly high cost.

Carbamates and Organophosphates Replace DDT

Amid these problems, biologist Rachel Carson in 1962 published *Silent Spring*. In it she marshaled evidence that DDT harms birds, mammals, and the ecosystem. Carson's critique led Congress to ban the use of DDT in 1972, though the furor over DDT did not end the use of insecticides. The carbamates and organophosphates that replaced DDT complicated the lives of farmers, who had to take special care in the handing and mixing of these insecticides. These new insecticides, which have been found to be dangerous to the central nervous system, have often caused headaches and have even been fatal. The need for caution has prompted state departments of agriculture to require farmers to demonstrate competence in the use of insecticides and to earn a license attesting to proficiency. To many farmers the requirement of licensure is a burden and an instance of government interference in their lives.

Whereas the gasoline engine in all its manifestations on the farm required the farmer to be a businessman, the carbamates and organophosphates demanded that he be a chemist and biologist. He needed to know not only how to mix insecticides for safety but also how to apply them in a combination that would kill only those insects that plagued his crops. Gone were the heady days of DDT when farmers indiscriminately killed insects. These new insecticides amplified the old problem of insect resistance, requiring the farmer to know which combination of insecticides was most effective, just as a physician must know which combination of antibiotics to prescribe against resistant bacteria. The carbamates and organophosphates are specific not only to a species of insect but to its stage of growth: some kill larvae, whereas others kill adults. This specificity of toxicity required farmers to scout their fields for broods of insects to learn their stage of development and to apply insecticides accordingly. In addition to exacerbating the burden of debt and the uncertainty of commodity prices, modern agriculture calls on the farmer to be a field naturalist.

Since the advent of DDT and, later, carbamates and organophosphates, chemistry has reinforced the trend toward rural depopulation. Just as the tractor pushed labor off the farm, the airplane and air-blast sprayer and the new insecticides they applied have diminished the need for labor.

Moreover, the new insecticides were more expensive than the previous generation of insecticides and favored farmers who had capital rather than the small farmer, tenant, sharecropper, and landless laborer. As did the tractor, insecticides have thus sharpened the divide between farm owner and laborer. In reducing labor, insecticides have reinforced the trend toward part-time farming. Increasingly, farmers held jobs in a nearby town or city and farmed during the weekend.

Farming without Insecticides

By the end of the 20th century, however, a small number of farmers had tapped into the dissatisfaction of some consumers with food from farms that used insecticides. These were the organic farmers who grew fruits and vegetables for local consumption and who minimized the influence of modern chemistry on their lives. Their daily routine harked back to the era of intensive farming and daily contact with plants.

The New Fertilizers

Parallel to the development of insecticides was that of new fertilizers. Farmers focused much of their attention on nitrogenous fertilizers because nitrogen more than any other nutrient affects the yield of crops. In the 1910s two German chemists devised a method, the Haber-Bosch process, of synthesizing nitrogenous fertilizers more cheaply than in the past, increasing their use on American farms. In the 1960s farmers began using anhydrous ammonia, a gaseous ion of nitrogen that fertilizer spreaders injected into the soil, as well as two other derivatives of ammonia: the solid ammonium nitrate and the liquid aqua ammonia.

The use of these new fertilizers transformed the lives of farmers. Gone were the days when farmers were generalists who grew several crops in rotation. Now able to get nitrogen from fertilizer they no longer needed to grow grains in rotation with alfalfa or another legume to fix nitrogen in the soil. Farmers now specialized in a single crop or a small number of crops: corn or corn and soybeans in the Midwest and wheat in the West. Without alfalfa or clover or another forage, farmers no longer had an incentive to keep livestock on the farm. The disappearance of livestock from the grain farm freed children from the routine of animal care, allowing them, as was the case with the tractor, to devote more time to school and social activities. Fertilizers thus reinforced the tractor in fashioning the lives of rural youth to mirror the daily life of suburban and urban youth. In diminishing labor on the farm, fertilizers, as was true of the tractor, hastened the move of people to the city. Moreover, fertilizers allowed farmers to simplify their lives. Without the necessity of growing legumes to fix nitrogen in the soil, farmers no longer needed barns in which to store forage or a mower and raker to cut and gather it.

Fertilizers also freed farmers for other pursuits. No longer beholden to care for livestock, farmers could work in a nearby town or city. Thus, fertilizers, like tractors and insecticides, reinforced the trend toward part-time farming. Moreover, farmers who had no livestock were freer than they had been in the past to vacation when they wished, particularly during winter, when they had more leisure than during the growing season. The ease of operation that fertilizers afforded farmers allowed people to farm who had no connection to the land and whose primary resource was capital. In 2000 people living at least two counties from their farm and who did not farm on a daily basis owned 39 percent of farmland in Iowa.

In diminishing the acreage devoted to forage and the number of livestock, fertilizers changed the working lives of farmers by increasing the imperative that they market their crops. When they had fed forage to livestock, they did not fret over the need to sell that portion of their crops. The transition from forage to corn or wheat monoculture integrated farmers into the global market. They now had to attend to commodity prices and to manage debt as the profit margin for these commodities narrowed. As had the tractor, fertilizers demanded that the farmer devote his time to business rather than to tending plants.

Fertilizers, like insecticides, required that farmers be chemists. Though cheaper because of the Haber-Bosch process, fertilizers were still too expensive to apply indiscriminately and so obliged farmers to test the soil to identify the nutrients in shortage and to apply only those fertilizers that supplied these nutrients. Gone were the days when farmers applied manure liberally to the soil. Without livestock, and faced with the high price of fertilizers, they have become frugal in fertilizing the soil.

Fertilizers, like carbamates and organophosphates, required farmers to exercise greater caution in their daily lives than they had in the past. The ammoniacal fertilizers irritate eyes and skin, obliging farmers to avoid contact with them. As a precaution, the farmer began keeping water with him while handling these fertilizers to wash them off should he inadvertently get them on his skin. Exposure to their fumes caused headaches.

Science and Technology Changed Rural Life

In an enormous variety of ways, then, science and technology influenced the lives of farmers in the 20th century. In the process, rural life changed in ways that Americans could scarcely have anticipated in 1900. At the beginning of the 20th century, half of all Americans earned their livelihood by farming, but by 2000 less than 2 percent did. People who had once worked on the farm moved to the city as the United States became an urban nation. Once a way of life, agriculture by 2000 had become a curiosity, an activity separate from the lives of most Americans.

Science and technology shaped agriculture in ways that would have been hard to predict in 1900. The demonstration of heterosis in corn in

1909 led to efforts to breed other crops for hybrid vigor. Parallel to the work with corn was the breeding of sugarcane, which agronomists had begun to hybridize as early as 1904. By the 1970s sugarcane hybrids yielded four times more than the old varieties. Since the 1960s the cultivation of the best hybrids in Florida expanded acreage at the expense of growers in Puerto Rico. Jamaicans, having left their homes for Florida, cut cane by hand, the labor-intensive method that had marked the harvest of sugarcane since the 16th century. Elsewhere in the United States the tractor, the combine, and other machines had the opposite effect, diminishing the need for labor and pushing workers from the countryside to the city in the 20th century—another contribution to the current number of 2 percent earning their livelihood by farming. In addition to sugarcane, hybrid varieties of cotton spread through the South and hybrids of wheat covered the Plains. The yield of hybrids rose rapidly with the application of fertilizers, and the use of fertilizer banished the old fear of American agriculture in the 19th and early 20th century: soil exhaustion.

So fruitful were the application of science and technology to American agriculture that they stood Malthus's ideas on their head. British cleric Thomas Malthus had supposed that human population would outrun its food supply, leading to hunger and misery for the masses. But the production of food on American farms kept pace with and perhaps even exceeded the growth in population in the United States. The large supply of food benefited consumers with low prices, but these cheap prices cut into farm profits. Science and technology taught the American farmer how to grow more food but not how to reap profits in an era of food surplus. To its credit, American agriculture feeds not merely Americans but people throughout the world with its surplus.

But Americans exported more than food to the rest of the world. They planted science and technology on foreign soil in the Green Revolution. High-yielding varieties of wheat turned India in the 1970s from a food importer to an exporter. Other countries in Asia, North America, Central America, South America, and the Caribbean planted high-yielding crops. As in the United States, farmers abroad combined these crops with fertilizers to double and triple yields. Agricultural science and technology have made possible the habitation of earth by more than 6 billion people. Whether the planet can sustain a geometric increase in the population into the future remains open to question.

2

TRANSPORTATION

Americans experienced a revolution in transportation in the 20th century. The electric railway, a technology of the 19th century, expanded from its base in the large cities to link cities in the Northeast, Midwest, and West in a network of track. More far-reaching was the automobile, which became so widespread during the 20th century that it touched nearly every facet of daily life. At the beginning of the 20th century, the railroad spanned long distances, but during the century the airplane rivaled and then surpassed the railroad as the carrier of passengers over the vast expanses of the United States.

THE ELECTRIC RAILWAY

A Romantic symbol of the 19th century, the locomotive spanned long distance, though it offered too few lines and made too few stops to service the jaunt from home to work or from home to store that made up the routine of daily life. Instead, in the first years of the 20th century and before the automobile took hold, Americans got from place to place locally on foot, by bicycle, and by horse-drawn wagon. An alternative to the old standbys of human and animal locomotion and one suited to everyday transit was the electric railway, or trolley. An electric railway that ran within a city was known as a street railway, and one that linked two cities was known as an interurban. An electric railway used an electric motor to generate motion. The danger of electrocution precluded the use of the track or a third rail to conduct electricity. Instead, a wire above the track carried electricity, which a pole on the roof of a railway car transferred to the motor.

By 1900 the electric railway was well established in the Northeast, Midwest, and Pacific coast. Ohio alone had 3,000 miles of track, and every city in the Buckeye State with more than 10,000 people had at least one line. In Indiana all but three cities with more than 5,000 inhabitants had railway service. During the first decade of the 20th century, interurbans extended into the countryside, linking farm and city. The most ambitious lines joined cities over vast expanses. By 1910 one could travel by interurban all but 187 of the 1,143 miles between New York City and Chicago and the entire distance between Cleveland and Detroit, between Indianapolis and Louisville, between Buffalo and Erie, and between Cincinnati and Toledo.

The electric railway gave Americans unprecedented mobility. Some lines ran at all hours of the day and night. In some cities trolleys ran every half hour. Seldom anywhere was the wait more than one hour between rides. Newspapers carried the schedule of rides, and to board a trolley one simply went to the nearest stop at the appointed time. At night a person who wished to board typically carried a newspaper with him, lit a match when he could see the lights of the trolley from a distance, set the newspaper afire, and waved it in the air to alert the driver as the trolley approached. A passenger could buy a ticket at a local store or, where one existed, at a station along the route. Often small and Spartan, these stations nonetheless sheltered passengers from the weather while they awaited a trolley. Fares were two cents per mile in the Midwest, with rates cheaper in New England and more expensive in the West.

Because of the distance they covered, interurbans offered passengers greater amenities than did the street railways, with sitting room, smoking room, and a toilet standard features. Some interurbans even had sleeping compartments, though these took up more space than many companies were willing to cede. Interurbans sold snacks, typically Crackerjacks, sandwiches, gum, and the like and even stopped for meals. Passengers who desired more carried their own food with them.

The electric railway changed rural life by easing farmers' isolation. In exchange for the right to lay track through their property, a company agreed to have a trolley pick up farmers along its route. Rural women took advantage of this convenience, using the trolley to run errands in a nearby town. The interurban increased the mobility of the housewife by including a compartment for her to store a bicycle. Once in town she shopped by bicycle, returning to the station in time for the trip home. While bringing mobility to the countryside, the trolley brought electricity to the farmer. A company that built the track through a farmer's land ran an electric line to his house and supplied him with electricity, typically for $1 a year.

The trolley also brought together college students, an important function given that many colleges were in the countryside and so separated by considerable distances. College sports benefited from the interurban in the way that professional baseball in New York City benefited from the subway. Oberlin College in Ohio once booked ten trolley cars to take 650 students and its football team to battle the Ohio State Buckeyes in Columbus. Baseball fans at both the collegiate and professional levels took the electric railway to games, secure in the knowledge that the same trolley would be waiting for them at the end of the game to return them home. An informal trolley league formed in the Midwest, with teams playing one another during summer. Amusement parks sprung up along trolley lines in the 1920s to entice pleasure seekers during the weekend. A creation of the Mellon family, Kennywood Park along a trolley line to Pittsburgh, boasted itself "the Nation's Greatest Picnic Park."[1] In bringing leisure to the masses, the electric railway took urbanites to the countryside to hike through the woods and pick berries and mushrooms.

Children, too, felt the effect of the electric railway, which shuttled them between home and school. This service was especially important in the countryside, where many children walked long distances to school. Alert to the potential of the electric railway to span distance, school officials in the early 20th century began to form school districts along trolley lines, designating as a single district those schools served by a single trolley. In this way the electric railway presaged the development of the school bus.

In addition to people, the electric railway carried the staples of daily life. Trolleys carried beer and newspapers to stores. Some were also part postal service—picking up mail, stamping it, and delivering it to post offices along their routes. Drivers doubled as errand boys, picking up prescription drugs and household items for riders during a stop.

The electric railway contracted after World War I. The mileage of track between cities declined from 15,580 in 1916 to 2,700 in 1939. The Pacific Electric Railway, the largest interurban in California, abandoned all but five lines in 1950 and sold these in 1954. In 1960 the Los Angeles Metropolitan Transit Authority converted the city's last interurban to a bus. In 1958 the Pacific Electric Power Company had abandoned its last and the country's oldest line, the route from Portland to Oregon City. Street railways suffered the same fate as the interurban. The number of streetcars in the United States declined from 56,980 to 4,730, and the mileage of track from 40,570 to 2,333 between 1929 and 1946.

THE AUTOMOBILE

During the 20th century, Americans gravitated to other modes of transit besides the electric railway: elevated train, subway, bus, and, most of all, the horseless carriage, or automobile. The elevated train, subway, and bus were creatures of the city, but the automobile roamed everywhere. The first automakers priced the automobile as a luxury: in 1901 the average car cost $1,600. Lansing, Michigan, automaker Ransom Olds sought to penetrate the middle class and priced his Oldsmobile that year at $650, roughly 40 percent of the average. Yet even this price exceeded the annual income of the average American. Henry Ford, who founded Ford Motor Company in 1903, did no better at first. The first Model A sold for $850 in 1903, the Model B for $2,000 in 1905, and the Model N for $600 in 1906. That year, Ford produced 10,000 Model Ns to meet demand and in 1908 unveiled the Model T. Although its initial price of $850 made the Model T more expensive than the Model N, with this line of automobile Ford crystallized his aim of mass-producing a standard car inexpensive enough that every American could afford it. In his quest to democratize transportation, Ford drove down the Model T's price to $490 in 1914, to $325 in 1916, and to $310 in 1921.

Ford reduced the cost of the Model T by making every phase of production efficient. In 1909 he announced that consumers could choose a Model T in any color as long as they wanted black, which he selected because it dried fastest. Moreover, concentration on a single color of paint reduced inventory and thereby cut the cost of storage.

Ford's obsession with efficiency and cost transformed the working lives of Americans. In 1910 Ford concentrated the manufacture of the Model T in a single factory at Highland Park, Michigan, where in 1913 he introduced the assembly line. Rather than shuttle workers from place to place, Ford moved automobiles in various stages of completion to them. Workers specialized in a small number of tasks that they repeated ad nauseam. Rather than the twenty minutes a single worker needed to build a magneto, a team of workers, each completing a single task, could churn out magnetos at the rate of one every five minutes per worker. Wherever possible, Ford mechanized production, making workers less craftsmen than machine operators, just as the tractor made the farmer less a tender of plants than a machine operator. As a result, the assembly line and its machines set the pace of work. After touring Highland Park, Charlie Chaplin would distill in the movie *Modern Times* (1936) the drudgery of the assembly line. Like the tractor, the assembly line never slackened its pace, and it offered workers only a fifteen-minute midshift break for lunch and use of the bathroom. Workers unaccustomed to this unrelenting pace and the logic of the assembly line that valued workers more as interchangeable parts than as humans felt alienated at Highland Park. So high was turnover in 1913 that Ford hired 52,000 workers just to maintain a workforce of 13,600. As interchangeable parts, workers could be discarded. Automakers and other businesses that employed factory workers favored workers in their twenties and early thirties; once in their forties, workers were in a precarious position. As their pace slackened, they became less valuable to an employer, and once laid off they had trouble finding a new job. In contrast, skilled workers—automobile painters, for example—had greater job security and longevity.

To create a compliant labor force amid this turnover, Ford hired Americans that other employers shunned: blacks, Latinos, ex-convicts, and the disabled whose disability did not impair productivity. Ford also departed from the tradition of paying a subsistence wage, raising wages to $5 a day, double the prevailing rate, as well as reducing the workday from nine to eight hours in 1914. Yet workers had to toil six months on probation before being eligible for $5 a day, and even then this pay was not automatic. Ford established a sociology department to enforce his ideal of clean living at home as well as at work. Only workers who lived up to the values of diligence, self-reliance, and frugality that Henry Ford had prized since childhood made $5 a day. Factory workers in other enterprises benefited from Ford's higher wages as other employers raised pay to compete. In 1919 Ford raised pay still further to $6 a day, though even at this rate Ford fell behind other automakers in the 1920s. By paying workers more than a subsistence wage, Ford and its competitors created a class of consumers able to afford automobiles, vacuum cleaners, washing machines, radios, and other amenities of life in 20th century America, which in turn increased profits for these companies. As it had with the Tracford and Fordson tractors, Ford touted the Model T's simplicity, durability, and ease of operation. Without frills or ornamentation, the Model T made few concessions to comfort. It was open, providing no protection against rain or snow, and one had to start the Model T by winding a crank rather than by turning an electric switch, endangering the operator. Two of the 3,100 skeletons in the Hamann-Todd Osteological Collection at the Cleveland Museum of Natural History in Cleveland, Ohio, have humeri that had once been broken. Anthropologists Kevin Jones-Kern and Bruce Latimer call these "Chauffeurs Fractures" to underscore their cause: the Model T's crank had slipped from the hand, recoiling against the upper arm to break the humerus.[2]

Such trauma hinted at the perils of driving. Before 1922, drivers operated the brakes by pulling a lever. Its use required strength and took more time than foot-operated brakes. A car needed plenty of space to stop. Worse, the brakes tended to wear unevenly, lurching a car to one side. Compounding this problem was the fact that automakers built cars for speed and power rather than safety. Poor roads and inexperienced drivers added to the hazards of driving. The result was a death toll in the early 20th century that exceeds current levels. Whereas the number of deaths was tallied at 1.7 per 100 miles traveled in 1995, the number was 24 per 100 miles in 1921.

Amid this carnage, states sought to impose order by posting speed limits and licensing cars and drivers. As early as 1901, New York required owners to register their cars. By 1910 thirty-six states, and by 1921 all states, required vehicle registration. By 1909 twelve states and the District of Columbia obliged drivers to have a license, and in the 1930s states began to require drivers to pass an examination to qualify for a license.

The Model T was adequate for plain folks, but Americans who wanted something more gravitated toward General Motors, which Detroit businessman William C. Durant had founded in 1908, the year Ford introduced the Model T. Rather than churn out a single model year after year as Ford did, General Motors offered several models in a range of prices. General Motors aimed to lure drivers by changing the appearance of its cars each year, making it easy for Americans to deride last year's style as outmoded. General Motors hoped not merely to coax Americans to buy a new car sooner than they might otherwise have but also to entice them to buy a more expensive model. Salesmen worked to get Chevrolet owners into a Pontiac, Pontiac owners into an Oldsmobile, and the owners of other models likewise to upgrade and in this way move Americans up the scale of ornamentation and price.

In its approach to sales, General Motors transformed the buying habits of Americans. Whereas Ford required buyers to pay in full for a Model T, General Motors pioneered the sale of cars on credit, creating in 1919 the General Motors Acceptance Corporation. Free from the need to scrape together the full price of a car, Americans could now buy cars that were otherwise beyond their means. By 1921 half of all buyers, and by 1926 three-quarters, financed the purchase of a car on credit. With the availability of credit, shoppers no longer needed to focus on buying a car for the lowest price and holding it for years. For a monthly payment, buyers could indulge their preference for amenities like electric starters, a closed carriage, or a color of paint and, when General Motors unveiled a new model, trade in their old car. Rather than coming to a dealer with cash in hand, buyers offered their old model as the down payment. A dealer in turn resold used cars at a fraction of their original value, attracting buyers who had neither the money nor the desire to buy a new car.

To stoke Americans' desire for cars, General Motors turned to advertising. No longer was it enough for an automaker to announce baldly the sale of a new car. General Motors and its imitators used large, illustrated ads to catch the eye and fire the imagination. Through its ads, an automaker sent prospective buyers the message that a car was not simply a means of transportation but a symbol of status or seductiveness. Attractive white women adorned the ads of cars to underscore the sex appeal of the automobile. Thanks to ads, the car evolved from a commodity to an amplification of its owner's personality.

Americans were receptive to these entreaties. U.S. automakers had kept pace with demand by manufacturing 1.5 million cars in 1921, but they built 4.8 million in 1929. Between these years, the number of automobiles registered in the United States rose from 10.5 million to 26.5 million, and by 1929 half of all American families owned a car, a fraction that England and Italy would not reach until 1970.

The automobile allowed Americans to live farther from work than had been possible in the era of the horse. Those who fled the cities in their cars settled in the suburbs, spurring their growth. The suburbs afforded their residents a larger yard than had been available in the city and so gave children more room for play and adults an area for a garden. Suburbs attracted Americans of similar income and ethnicity, and in this way the automobile fostered homogeneity.

The mobility that led Americans to the suburbs shaped even the lives of those who did not own a car. In the early 20th century, when physicians still made house calls, a person who required care needed only to phone his physician. Among the first buyers of cars, physicians were able to make house calls with a speed and at a distance that had been unthinkable in the era of the horse. In facilitating access to medical care, the automobile was especially important in the countryside, where physicians had few alternatives to the automobile.

Physicians were not alone in integrating the automobile into their lives. Criminals used it to rob banks and during Prohibition to transport alcohol. Alert to the new mobility of criminals, police adopted the squad car, reshaping the nature of police work. Less common were the police officers on foot patrol who were a fixture in a neighborhood, able to talk familiarly with people while on their rounds. The automobile put the police in traffic rather than on the sidewalk, limiting their interaction with people.

Aside from its role in transportation and work, the automobile reshaped leisure much as Henry Ford had foreseen in 1908. Families went for Sunday drives after church, a practice that eroded the blue laws. The automobile freed Americans to vacation farther from home than people generally dared to travel in the era of the horse. Americans took to the road, visiting national monuments and historic places. Property owners near these sites built or enlarged hotels to accommodate vacationers. The automobile thus spurred the growth of tourism in the United States.

Adults were not alone in gravitating toward the automobile. Adolescents used it to expand their social lives. On occasion, the automobile led to quarrels as children sought greater freedom in using the family car than parents were willing to concede. One source of tension was the use of the automobile in dating. Young couples, chaffing under the convention of visiting each other under the watchful eye of parents, took their courtship to the road, using the car to go parking. Free from the supervision of parents, couples could explore their sexuality, bringing to fruition the message of ad makers that the automobile gave its driver sex appeal.

Driving on this scale required infrastructure. The petroleum industry expanded to meet the demand for gasoline, a distillate of petroleum. The first gas station opened in 1905, and in the 1910s Standard Oil and its competitors began to fill the United States with a network of gas stations. In the 1920s the discovery of petroleum in Louisiana, Oklahoma, and Texas expanded the supply of gasoline faster than demand and kept the price of gasoline low. Americans enjoyed cheap gasoline until the 1970s and came to regard this circumstance as the natural state of affairs. Automakers built cars for speed and power rather than for fuel economy. The oil boom in the American South and Southwest created new jobs related to the extraction, refinement, and transport of petroleum and generated wealth for a region that had been in the doldrums since the decline of sugar culture in Louisiana after the Civil War.

Roads were a second component of infrastructure. Before the automobile, only cities had asphalt, stone, or brick roads. Americans in the countryside made do with dirt roads that were rutted from wagon traffic and that turned to mud in the rain. In 1900 the United States had only 200 miles of paved roads in the countryside. In 1908, the year Ford introduced the Model T, Wayne County, Michigan, began to pour the nation's first concrete road. World War I, in requiring the United States to move soldiers and materiel across the country, underscored the value of good roads. Americans had long considered the building of roads the responsibility of localities and states, but the war at last prodded the federal government to act, and in 1916 Congress spent $75 million on the construction of new roads and the paving of old. In 1921 Congress funded the construction of what would become the first hard surface road linking the Atlantic and Pacific coasts, U.S. Route 30, known to many people as the Lincoln Highway. By 1929 the United States had 700,000 miles of hard surface roads, and the states and Congress were spending some $2.5 billion a year on road building. During the Great Depression, Congress funded the building of roads as a way of putting people back to work, and in 1940 California opened the Pasadena Freeway in Los Angeles, the nation's first high-speed highway. As had World War I, World War II heightened the need for more and better roads, and during the war Congress funded 75 percent of the construction of new roads. Dwight D. Eisenhower, a hero of World War II, recalled as a young officer traveling across the United States on a maze of poor roads. As president he determined to span the United States with a network of interstate highways, and at his urging Congress in 1956 passed the Federal-Aid Highway Act giving the states $25 billion to build 41,000 miles of highway. Federal spending on roads leapt from $79 million in 1946 to $2.9 billion in 1960 and to $4.6 billion in 1970.

Highways transformed the American landscape. The narrow, sometimes winding roads of the early 20th century ceded ground to vast expanses of multilane, straight highways. Motoring along these highways for long durations introduced to driving a degree of monotony that had been unknown at the beginning of the 20th century. Highways wrought economic changes as well. The towns along the old roads stagnated while new towns and cities expanded along the highways. For example, Interstate 70 brought Junction City, Kansas, six new motels and several restaurants. The highways blurred the distinction between cities, and in the 1960s urban planners coined the term *megalopolis* to describe the sprawl of cities between Boston and Washington, D.C. Highways and the megalopolises they spawned also spread from Pittsburgh to Milwaukee and from San Francisco to San Diego.

Improvements in technology aided Americans as they took to the highways. In 1966 Oldsmobile designed the Toronado, the first American car with front rather than rear wheel drive, an innovation that improved the handling of a car. That decade, automakers converted many models from manual to automatic transmission and introduced power brakes and steering and air conditioning.

The automobile exalted both family and individual. In 1965 Ford converted its Mustang from two to four seats, making room for two people in the back, and touted the Mustang's versatility. A mother could run her children to school and shop, and a father could, in his off hours, zoom around town in something akin to a sports car. The Mustang was thus both family car and a symbol of the individual's longing for speed and power.

Postwar prosperity and the ubiquity of the automobile accelerated the growth of suburbs. By 1970 the majority of jobs in nine of the country's fifteen largest cities were in the suburb rather than in the incorporated area, and people who did work in the city often lived in a suburb of it. In 1970 people who lived in the suburbs accounted for three-quarters of all trips to work in San Francisco. Increasingly, people came and went by car as the number of rides on public transportation in the United States fell from 23 billion in 1945 to 8 billion in 1967.

Increasingly, too, Americans sat behind the wheel of an import. In 1955 foreign cars accounted for only 0.5 percent of U.S. auto sales, but in 1965 Toyota began selling its Corona in the United States. In contrast to American automobiles' prodigal use of gasoline, Toyota touted the efficiency of the Corona. Families turned to the Corona or another import as their second car, a no-frills model for errands and transporting children to Little League. The drift toward imports quickened after 1973, when the Organization of Petroleum Exporting Countries halted shipment of petroleum to the United States to protest U.S. support of Israel in the Yom Kippur War. The spike in gasoline prices and calls for rationing led people to line up outside gas stations in near panic. Americans who were unable or unwilling to use public transportation instead bought fuel-efficient foreign cars. By 1975 imports accounted for nearly 20 percent of U.S. auto sales.

American automakers struggled to remain afloat. By 1980 General Motors owned 35 percent of Japanese automaker Isuzu and fought to regain market share with the Vega and other compacts. Ford entered the market of fuel-efficient cars with the Pinto, and American Motors the Gremlin. In the 1990s Ford marketed a line of compacts and sales of its light truck the Ranger outpaced those of the F-150. Likewise, General Motors had the S-10 and Dodge the Dakota as counterweights to their larger line of trucks.

The vagaries of gasoline prices did not turn Americans away from the automobile, however. By the end of the 20th century, it had firmly ensconced itself in American life. Americans could buy a Big Mac or Happy Meal at a McDonald's drive-through. Banks, pharmacies, and dry cleaners encouraged people to do business from their car. Even some funeral homes allowed mourners to pay their respects by car, and some chapels married couples in their cars. Americans had integrated the automobile into their lives in ways that no one could have envisioned in 1900.

THE AIRPLANE

In 1903, the year Henry Ford founded Ford Motor Company, Wilbur and Orville Wright took flight in Kitty Hawk, North Carolina, ushering in the aerospace age, though the airplane remained a curiosity and a tool of warfare rather than a commercial venture for the next two decades. In 1914 the first American airline to carry passengers flew between Tampa and St. Petersburg, Florida, but the company's failure left the fledgling carriers to question the profitability of passenger service. Publisher, aircraft enthusiast, and early advocate of air travel Alfred Lawson aimed to build an airline, but his first flight crashed in 1921, and Lawson abandoned hope. More successful in the short term was Aeromarine West Indies Airways, which the previous December had begun flying passengers from Key West, Florida, to Havana, Cuba. The success of this route led Aeromarine to expand service to New York and the Great Lakes, and in 1921 and 1922 the airline flew nearly 750,000 passengers to destinations in the United States and the Caribbean. In 1922 Aeromarine began twice-daily service between Detroit, Michigan, and Cleveland, Ohio, though the $25 fare one way nearly tripled the $9 by train. Never very profitable, Aeromarine went out of business in 1924. As if to reassure carriers after Aeromarine's demise, the National Aeronautics Association's Aeronautical Review in December 1924 proclaimed the soundness of air travel as a "business enterprise."[3]

More enduring in the 1920s were the firms that carried mail as well as passengers. With room for as many as sixteen passengers and for hostesses who served lunches and snacks, the carriers began to attract businessmen and women who saved time by flying rather than taking a train or interurban. Transcontinental Air Transport inaugurated coast-to-coast service on June 20, 1929, covering the route between New York City and Los Angeles partly by plane and partly by train, a trip that took two days and cost $337 one way. The portion of the ticket for the plane flight alone cost $290, roughly 16 cents per mile, eight times the rate for travel by interurban. High as they were, these rates did not deter Americans from flying. The number of passengers leapt from 5,782 in 1926 to 172,405 in 1929.

In the 1920s the commercial airlines used engines left over from World War I. These were powerful but needed maintenance every fifty hours of flight time. Such frequency made the maintenance of engines half the cost of operating an airplane. This would not do, and airlines pressed the manufacturers of engines to upgrade quality. Early versions of the Pratt and Whitney Wasp engine needed maintenance every 150 hours. By 1929 most engines needed maintenance only every 300 hours. By 1936 engines had stretched maintenance to every 500 hours and the cost of maintenance had dropped by 80 percent. Not only were aircraft engines more durable, they were more powerful. Between 1930 and 1939, the Wright Cyclone engine doubled horsepower from 550 to 1,100.

Coaxed into the manufacture of airplanes by son Edsel, Henry Ford decided to acquire an airplane manufacturer rather than build a model from scratch. In 1925 Ford bought Stout Metal Airplane Company, shunting aside founder William Stout, who left the company. In 1928 Ford began manufacturing the Ford Tri-Motor on an assembly line, as he was doing in manufacturing automobiles. Like the Model T, Ford designed the Tri-Motor to be durable and Spartan. In keeping with its name, the Tri-Motor had three engines, each a 220 horsepower Wright j-5 air-cooled radial engine. The Tri-Motor had a cruising speed of 122 miles per hour and could climb to 18,500 feet. It sat fifteen passengers in wicker chairs and had a washroom in the rear. Ceasing production in 1933, Ford had built 200 Tri-Motors and had sold them to more than 100 airlines.

Flying was not always a pleasant experience in its early years. Temperatures in a poorly heated cabin could drop below 50 degrees Fahrenheit, and passengers had little choice but to accept noise and vibration as the price of speed. The Tri-Motor flew low in the sky by today's standards and was subject to air that was more turbulent than at higher altitudes. The air in the passengers' cabin was poorly ventilated, and into this air leaked engine exhaust and fumes from engine fuel. Passengers could count on the copilot for aspirin, but in turbulent weather aspirin was not always enough. Each passenger seat came with bags and a box in the event a passenger became sick, but the passenger sometimes opted instead to vomit out the window. Some passengers did not go to the trouble of opening the window and vomited in the cabin instead. Some passengers carried mail on their laps, so short of space were the planes. Unable to talk over the roar of the engine, passengers resorted to passing notes to one another. Alert to these problems, the Douglas Aircraft Company hired Sperry Rand Corporation to improve comfort. Sperry engineers outfitted the DC-1 with soundproof walls that lowered noise by 25 percent and mounted seats with upholstering and armrests on rubber platforms to absorb vibration. Douglas boasted that its DC-4E had a cabin larger than the coach of a train, full soundproofing, steam heat and air conditioning, hot and cold running water, bathrooms large enough for a passenger to change clothes, and electric outlets. Families gathered around a table for meals as though they were at a restaurant as air travel came in the 1930s, despite the Great Depression, to attract Americans who flew for pleasure as well as for business, much as the passengers of Aeromarine had gone to Cuba for sun and alcohol in the era of Prohibition.

As airlines moved away from the austerity of their early years, they hired stewardesses to serve coffee and food. Western Air Express hired fifteen stewardesses in 1928, and United Air Lines hired eight in 1930. The airlines required their stewardesses to be registered nurses and, perhaps more appealing from the perspective of male passengers, screened them for age, height, weight, and attractiveness.

Rather than retrenchment, the Great Depression witnessed the building of new airplanes. In 1933, the year Ford ceased production of the Tri-Motor, Boeing began manufacturing its 247. It had two Pratt and Whitney air-cooled radial engines and carried ten passengers and a cadre of stewardesses. Like the Tri-Motor, the 247 had a washroom for passengers. Boeing sold its 247 to United Airlines. The tie between Boeing and United was intimate. United Aircraft and Transport Corporation owned both Boeing and United. Being in business together, United bought whatever airplanes Boeing made. By June 1933 United had thirty 247s, putting them to work on the ten daily round trips between New York City and Chicago.

Determined to have a better airplane than the 247, TWA contracted Donald Douglass to build a bigger, faster plane. The result was the DC-1, which could hold twelve passengers, compared with the ten held by the 247. The DC-1 made its inaugural flight in 1933 but never went into production because Douglass was already at work supplanting it with the DC-2 in 1934 and the DC-3 in 1936. The biggest of the trio, the DC 3, could hold twenty-one passengers and was also faster than the 247. So superior was the DC-3 that United bought them to replace its fleet of 247s. In 1936 American Airlines put the DC-3 in competition with the 247 on its New York City to Chicago route.

After struggling during the first years of the Great Depression, Lockheed revived its fortunes in 1934 by building its Model 10 Electra to compete with the 247 and the DC series. The Electra seated ten passengers in a soundproof cabin. Its two 450 horsepower Pratt and Whitney Wasp Junior air-cooled radial engines gave the Electra a top speed of 206 miles per hour, faster than the 247 and the DC line, and a ceiling of 21,650 feet. Lockheed built 149 Electras, selling most of them to Northwest Airlines and Pan American.

Thanks to new levels of comfort and service, flight became an experience in its own right rather than merely a means of shuttling people between locations. In 1929 airlines experimented with showing motion pictures to passengers, and in 1934 Transcontinental and Western Air made cinema standard on its flights. After World War II, passengers could choose from among a variety of entertainment and information: movies, taped music, news or comedy, current issues of newspapers and magazines, and, under the direction of the pilot, they could even dial in a program broadcast from a nearby radio station. Airlines served wine, liquor, and mixed drinks as well as a more sobering choice of coffee, tea, or soda. Bars, lounges, and even pianos became the foci for socializing. Smokers had their own section of the plane, usually in the back, and could choose among brands of cigarettes during a flight. The advent of frozen foods allowed the airlines around 1950 to expand meals beyond cold cuts and sandwiches.

Depending on preference and need, passengers could request kosher, vegetarian, and dietetic meals.

Despite advances in comfort and service, airlines could not alter the circadian rhythm, however. Passengers who flew across time zones began after World War II to complain of fatigue and difficulty concentrating for days after their flight. Even President Dwight D. Eisenhower's secretary of state, John Foster Dulles, complained that these effects had marred his negotiations with Egypt and were partly to blame for the Suez Crisis in 1956. In the late 1950s physicians coined the term *jet lag* for the phenomenon and counseled rest and a light schedule in the days after a flight, a regime ill suited for the business traveler who faced a round of meetings.

Americans, impervious to jet lag or willing to suffer its effects, took to the skies on vacation. Before the aerospace age, families had vacationed, when they did so at all, near home. The train and automobile made possible long trips, but these took days. The airplane brought to a broad cross section of Americans the experience of roaming the country or even the world. Airline ads lured them with the promise of sun in Miami or casinos in Las Vegas. The airplane made possible the weekend jaunt to Paris or Rome and so opened a new era in vacationing. No longer were exotic locales the privilege of the wealthy.

More mundane was the experience at an airport. With its retinue of restaurants and hotels and separate from the city it served, an airport was a self-contained unit. A business executive who flew into O'Hare International Airport in Chicago, Illinois, for example, could meet a client, eat his meals, and sleep in a nearby hotel without ever venturing into the heart of Chicago. In this way airports promoted an insularity that countered the airplane's capacity to bring people together.

The airplane of the postwar period was the Boeing 747. In 1969, Boeing began manufacturing the plane, which was for years the signature aircraft of Pan American. The 747 had four 43,000-pound Pratt and Whitney JT9D turbofan engines, and it carried 490 passengers, had a cruising speed of .85 Mach, a cruising altitude of 34,700 feet, and a range of 6,828 miles, making possible nonstop flights from the United States to destinations in Europe and Asia.

Rising to challenge Boeing were Lockheed and McDonnell Douglas. In 1970 Lockheed began manufacturing the L-1011 Tristar, which had three Rolls Royce RB 211 jet engines and could carry 400 passengers. That year, McDonnell Douglas unveiled its DC-10, which had, depending on its specifications, three General Electric CF 6 turbofan engines or three Pratt and Whitney JT9D turbofans. Neither the Tristar nor the DC-10 could carry as many passengers as, fly as fast as, or cover as much distance as the 747.

As Henry Ford had earlier done to the automobile, airlines in the 1970s began to price tickets within the means of nearly every American. The first bargain airline, Southwest, began flying in 1971 at a fare of $20 one way for all its routes and $13 evenings and weekends. In 1973 Southwest drove

This Boeing 747 has just gone airborne. First manufactured in 1969, Boeing had built some 1,400 Jumbo Jets by 2006. Courtesy: Shutterstock.

down its fare to $13 and introduced the principle of the pit stop to air travel, aiming to take only ten minutes to disembark one set of passengers and load the next set. Southwest skimped on food and legroom, replacing meals with coffee, juice, and peanuts and packing seats on its planes.

Since the 1970s, air travel, although part of the daily lives of Americans, has become hectic. The increase in the number of passengers that Southwest and its imitators made possible with their low fares produced long lines at airport terminals. Delays and cancelled flights, more common as airlines offered more flights, marred the experience of flying for vacationers and businesspeople alike. As with the Model T, flying became less a memorable event than a way of getting from place to place as cheaply and quickly as possible.

The bargain airlines shaped the working lives of their employees. Casual dress prevailed at Southwest, as did Friday deck parties in Dallas, golf outings, and cookouts. Eager to underscore the sex appeal of its flight attendants (stewardesses), Southwest dressed them in hot pants, leather boots, and formfitting tops. Its first chief hostess had worked aboard Hugh Hefner's jet. Rachel Welch, quipped one barb, would not measure up at Southwest. Free from unions, Southwest could dictate wages, paying sales agents $8.05 an hour in 2000 and its flight attendants, working on a piece rate, averaging $8.67 an hour.

The first U.S. airline to host a Web page, Southwest encouraged Americans to buy their ticket online. Other airlines followed, requiring passengers to buy their ticket online to capture advertised special prices. Airlines encouraged passengers to sign up for a weekly e-mail of all specials. Travelocity.com and other search engines compared fares among airlines for a specified date. Yet in 2000 only one-third of Americans booked their flight online, compared with more than 80 percent in the United Kingdom.

At the end of the 20th century, Americans still relied on the automobile to get from place to place. Popular as it was, the automobile nonetheless faced criticism. The OPEC oil embargo of the 1970s demonstrated how dependent the automobile was on petroleum imports. No less serious were the critics who pointed to the automobile as a polluter. Auto exhaust includes compounds of nitrogen and sulfur, which pollutes the atmosphere and returns to earth as acid rain. The carbon dioxide in automobile exhaust raised the level of carbon dioxide in the atmosphere and so contributed to global warming.

Searching for a cleaner alternative to the automobile and for a way of abating traffic congestion, some city planners resurrected the electric railway. This resurgence in the trolley reversed the trend toward the automobile that had brought the trolley near extinction. Portland, Oregon, and Saint Louis, Missouri, built trolley lines that attracted large numbers of riders. The Saint Louis line is particularly felicitous, as it follows the city's grid east and west. Buses run north and south off the trolley line. Although the electric railway emits no pollutants, representing an advance over the automobile, the power plants that supply electricity to the trolley burn coal or natural gas. Coal is the worse polluter, but even natural gas is imperfect. The burning of natural gas releases carbon dioxide into the atmosphere, exacerbating the problem of global warming.

The search for a clean form of transportation has no easy answer. During the oil crisis of the 1970s, fitness enthusiasts reminded Americans that the bicycle was a form of transportation. True, people in much of the world—China and India are examples—get from place to place by bicycle. It is also true that the bicycle is a clean technology, emitting no pollutants other than the minute amount of carbon dioxide that the rider exhales. One might imagine Americans within five miles of work talking a bicycle, and yet the ecological advantages of a bicycle have never been enough to convince Americans to part with their cars.

Perhaps the only way to get Americans off the highways is to get them to work from home. Some companies already permit staff to work from home, and if the trend expands to millions of workers, the practice of telecommuting could lessen Americans' dependence on the automobile and thereby reduce pollution. Telecommuting would end, or at least reduce, the peak hours of driving and thus reduce gridlock. Telecommuting would not, however, end the use of the automobile. Telecommuters would still retain a car for running errands, visiting the doctor, dropping off a book at the library, and a dozen other tasks. Ironically, many Americans could do most of these errands by bicycle, but they do not because they still prize their automobiles.

The reality, one suspects, is that most Americans will not give up their automobile, no matter how dire the environmental crisis. Change might come, however, in the form of what kinds of cars Americans buy—swapping a gasoline-powered car for, for example, a hybrid, which employs a rechargeable battery to propel the car at the low speeds of stop-and-go city driving. Homes clustered near mass transit lines may encourage many Americans to take public transportation. Tax credits could entice people to carpool. These ideas might persuade Americans to drive less than they do now and so to conserve fossil fuels.

NOTES

1. Ruth Cavin, *Trolleys: Riding and Remembering the Electric Interurban Railways* (New York: Hawthorn Books, 1976), 53.

2. Kevin Jones-Kern and Bruce Latimer, "History of the Hamann-Todd Osteological Collection," http://www.cmnh.org/collections/physanth/documents/Hamann Todd_Osteological_Collection.html.

3. *Aeronautical Review* (December 1924). Quoted in Kenneth Hudson and Julian Pettifer, *Diamonds in the Sky: A Social History of Air Travel* (London: British Broadcasting Corporation, 1979), 27.

3

HEALTH AND DISEASE

In 1900 medicine faced several challenges in the United States. Plague assailed the residents of San Francisco. The *Aedes aegypti* mosquito, and with it yellow fever, threatened New Orleans, as it had for more than a century. Hookworm and pellagra—a parasite and a disease, respectively—were endemic to the South, robbing their victims of vitality. Smoking was a dangerous pastime, particularly among men, though physicians were then unaware that tobacco caused cancer and heart disease. Childbirth was still a painful process, and there was then little science in the rearing of infants.

Medicine responded to these challenges. In 1900 U.S. Army officer and physician Walter Reed identified *Aedes aegypti* as the carrier of yellow fever. Working in Havana, Cuba, Reed directed the draining of standing water and, where this was impractical, the mixing of kerosene in standing water. Thus deprived of water in which to lay its eggs, *Aedes aegypti* died out in Havana. New Orleans was slow to implement these measures, and only in 1905, in the midst of an outbreak of yellow fever, did municipal authorities undertake to eradicate *Aedes aegypti*.

In 1902 zoologist Charles Stiles identified the parasite, a worm, that caused hookworm. The fact that the worm burrowed into the soles of the feet suggested a cure in the form of wearing of shoes. Pellagra, a nutritional deficiency, was a more difficult challenge. The solution lay in the discovery of vitamins, the discovery of which not only led to the cure of pellagra but also advanced nutrition as a science.

The case against tobacco lay not in a single discovery but rather in the accumulation of evidence that tobacco was a carcinogen and a cause of heart disease. Since the 1960s, the U.S. surgeon general has issued stern warnings against the smoking of tobacco. The campaign swelled amid a nationwide crusade to get Americans exercise to become fit and thereby reduce the risk of heart disease and some cancers.

Women led a movement in the early 20th century to combat the pain of childbirth with drugs. Other women insisted that drugs, in dulling the senses, robbed mothers of one of life's transcendent experiences. Many women breast-fed their infants but others looked to the new science of nutrition for vitamin-fortified formula.

INFECTIOUS DISEASES, VACCINES, SULFA DRUGS, AND ANTIBIOTICS

The bacterium *Yersinia pestis* causes bubonic plague. Fleas transmit plague through their bite. Feeding on rats, infected fleas transmit plague to them, making plague primarily a disease of rodents. But when a large population of infected fleas and rats lives near humans, the potential exists for either fleas or rats to infect humans by biting them. Symptoms of plague include fever, chills, headache, diarrhea, and swelling of lymph nodes. This last symptom gives bubonic plague its name, as buboes are the swollen areas of the body. In the era before antibiotics, roughly half those who contracted plague died from it. Plague is an ancient disease, though the term was used loosely, making difficult the identification of the disease in ancient sources. The outbreak of disease in 541 c.e. in Constantinople is the first that historians confidently identify as plague. The most fearsome outbreak occurred between 1347 and 1351, when plague swept Eurasia, killing perhaps one-third of the population. Plague revisited Europe in the 15th and 16th centuries.

Plague came to the United States from Asia. The telltale buboes marking its victims, plague struck Hawaii in January 1900 and San Francisco in March. The first victim in San Francisco, Chinese laborer Chick Gin, was found dead on March 6. Hesitant to admit the presence of contagion, newspapers and community leaders denied that plague had been the culprit. Brushing aside these denials, the city quarantined Chinatown in early March, but the plague did not abate, and it went on to kill three more Chinese Americans. Amid talk of halting Asians from fleeing San Francisco and sensational new reports comparing the outbreak in San Francisco to the plague that ravaged London in 1665, the plague retreated as quickly and silently as it had struck.

Plague reminded Americans in 1900 of what they would have preferred to forget: infectious diseases stalked them, carrying off the young and old. In 1900 influenza and pneumonia together killed more Americans than any other cause, taking 203.4 lives per 100,000 people. Tuberculosis ran second, at 201.9 deaths per 100,000, and as a group typhoid, smallpox, measles, scarlet fever, and whooping cough carried off 115.9 per 100,000.

In contrast, cancer killed only 63 Americans per 100,000, and heart disease fewer still. Americans in the early 20th century had more to fear from microbes than cholesterol.

As is often the case in human affairs, luck and hard work played their part in combining against the microbes of death. In the early 20th century, Americans stressed the importance of hygiene in the battle against pathogens, and nowhere was the mantra of hygiene repeated more often than against tuberculosis, the white plague. Before the discovery of a vaccine and the antibiotic streptomycin, physicians counseled fresh air and sunshine, good hygiene, and sanitary conditions as the bulwark against the spread of tuberculosis. These recommendations took on greater urgency in 1917, when a new test for tuberculosis revealed the disease in latent form in people. The test made evident that thousands of Americans who otherwise appeared healthy had tuberculosis. This news was alarming given the ease with which tuberculosis, a respiratory disease, spreads. The bacterium that causes tuberculosis settles in the lungs, from which it spreads through coughing. A person need not be in contact with someone infected with tuberculosis; proximity is enough to contract the disease. Antituberculosis societies were quick to act on the news that the disease was widespread in latent form and, in doing so, to affect the lives of Americans. As an adjunct to the Christmas seal campaign, antituberculosis societies targeted a hygiene crusade at America's children in hopes of instilling good habits early in life. The crusade reinforced the importance of basic hygiene, awarding children points for "hygiene chores."[1] More than a part of the daily routine, brushing one's teeth became an action against tuberculosis. By 1919 more than three million American children participated in the crusade.

Hygiene was not enough to combat infectious diseases, though. Strong measures were necessary. In this context, vaccination became one defense against disease as scientists in the 20th century developed vaccines against tetanus, tuberculosis, smallpox, measles, and whooping cough. Amid this silent revolution in the treatment of diseases came the sound and fury of the campaign against polio. Americans in the 1930s could point to President Franklin D. Roosevelt as a victim of polio. The National Foundation for Infantile Paralysis, tugging on the emotional ties that bound Americans to their president, campaigned on radio for people to send donations in dimes to the White House to mark Roosevelt's birthday on January 30, 1938. Americans responded with more than $260,000 in dimes. More heartrending than Roosevelt's misfortune was the fact that polio crippled the young, a circumstance that a child-centered America could not endure.

Eager for science to prevent the spread of polio, the foundation put its donations to the development of a vaccine. Two types of vaccines were created: one using a live or replicating strain of a microbe in a weakened state to stimulate an immune response in the body, and another using a dead or nonreplicating strain of a microbe. Unsure which type of vaccine

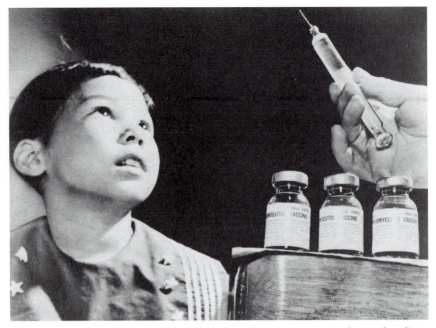

This boy awaits a polio vaccine. In the 20th century vaccines saved countless lives in the United States. Courtesy: Library of Congress.

would be more effective, the foundation funded research toward the development of both. University of Pittsburgh physician Jonas Salk headed the effort to develop a vaccine of nonreplicating poliovirus, and University of Cincinnati physician Albert Sabin, a vaccine of replicating poliovirus in a weakened state. Salk developed his vaccine first, and on April 12, 1955, ten years after Franklin Roosevelt's death, news that the Salk vaccine was effective and safe prompted Americans to ring church bells, throng the streets in impromptu celebrations, let children out of school early, and organize parades.

The scientists who developed vaccines trod a well-worn path, but the discovery that chemicals could kill pathogens in the body without harming a person carved a new trail through the lives of Americans in the 20th century. This fortuitous discovery stemmed from the practice of staining cultures of bacteria with dyes to highlight them under the microscope. Serendipitously, German chemist Paul Ehrlich discovered in 1906 a dye that killed the bacteria he was examining and, encouraged by this find, Ehrlich in 1910 developed Salvarsan, the first chemotherapy against bacteria. German physician Gerhard Domagk followed Ehrlich, discovering in 1935 that the red azo dye prontosil was toxic against streptococci. Scientists at the Pasteur Institute in Paris isolated sulfanilamide, the active ingredient in prontosil, opening the era of sulfa drugs.

Listening to Domagk describe the wonders of sulfanilamide at a conference, British physician Alexander Fleming believed his penicillin had the potential to outstrip sulfanilamide in fighting bacterial infections. Fleming had in the 1920s tried innumerable biological agents—he believed tears and egg whites held promise in killing bacteria—and so grew bacteria of many strains in petri dishes as a matter of course. In 1928 he chanced upon a single dish contaminated by the mold *Penicillium notatum*. Noting that the bacteria in the dish had died, Fleming surmised that the mold had produced a chemical toxic to them. He named the chemical penicillin to signify its derivation from *Penicillium* and classified it an antibiotic. Able neither to isolate penicillin in pure form nor to produce it in quantity, Fleming made no headway in treating bacterial infections with it. Others isolated penicillin, but the breakthrough in production came only in 1942.

The United States, a combatant in World War II, needed penicillin in bulk to treat its troops, and in a crash program similar to the more well-known Manhattan Project, began efforts to mass-produce the antibiotic at the U.S. Department of Agriculture's Northern Research Laboratory in Peoria, Illinois. Robert E. Coghill, director of the lab's fermentation division, began growing *Penicillium notatum*, Fleming's mold, in corn steep liquor and lactose. Not satisfied, although Coghill and his research team had increased the production of penicillin by more than twenty times over what any other method had achieved, a research associate happened upon a rotten cantaloupe in a Peoria grocery and isolated *Penicillium* chrysogenum, a strain that produced more penicillin than Fleming's mold had. The new production techniques and new species of *Penicillium* enabled the United States in 1943 to produce penicillin for its troops and those of Great Britain, whereas in 1942 the United States had had only enough for one hundred patients. The production of penicillin for civilian use began in 1945, and by 1958 the United States had 440 tons of penicillin, enough for every American to have two million units of the antibiotic. Parallel to the work of the Northern Research Laboratory was the effort of Rutgers University microbiologist Selman Waksman to discover new antibiotics. Aware that the tubercle bacillus does not grow in soil, Waksman and graduate student Albert Schatz screened soil microbes for the production of a chemical toxic to the bacterium, discovering in 1943 streptomycin, the first antibiotic effective against tuberculosis.

Antibiotics remade American life. In 1940 playwright and Nobel laureate Eugene O'Neill, himself a tuberculosis survivor, had cast the shadow of tuberculosis across the Tyrone family in *Long Day's Journey into Night*. A decade later, sanitariums were closing for lack of patients. The closing of the renowned Trudeau Sanitarium in 1954 made the front page of the New York *World Telegram* and confirmed for some physicians the status of streptomycin as a wonder drug. In penicillin Americans at last had a weapon against syphilis and gonorrhea, but some feared the antibiotic might make possible risk-free sex and thereby seduce men and women into promiscuity. After World War II, the army, alert to the inclinations of its recruits toward casual sex in the era of antibiotics, required officers to spend one hour each week "building character" in their charges.[2] Social commentators feared the demise of the family and blamed the sexual revolution of the 1960s on the cavalier attitude toward sex that penicillin had fostered.

During these years, antibiotics underwent a revolution. The first antibiotics could not pass through the stomach without losing their potency and so were injected into the body. Unable to self-administer injections, the patient surrendered treatment to the hospital, whose nurses monitored the dose and timing of injections. The development of oral antibiotics, however, allowed the patient to administer an antibiotic in the home. Away from the supervision of the hospital, less conscientious patients missed doses or even stopped taking an antibiotic when they began feeling better. Erratic and incomplete, this type of treatment allowed bacteria resistant to an antibiotic to develop and multiply, creating the problem of drug resistance that has beset Americans since the 1950s and that the rise of Acquired Immune Deficiency Syndrome has only worsened. Adding to this problem, Americans who visited their doctors for a complaint, no matter how minor, expected an antibiotic as a matter of course. Some parents were insistent that their child's pediatrician should prescribe an antibiotic for a stuffy nose or other innocuous complaint. Even when physicians suspected a viral infection, which is impervious to antibiotics, the temptation to placate the patient with an antibiotic too often overrode the reality that an antibiotic would do no good in these cases. The result was a culture in which Americans came to believe that a pill existed for every ailment and that the physician's primary job was to dispense medicine.

YELLOW FEVER IN LOUISIANA

The flavivirus yellow fever virus causes yellow fever. Comparing the protein coat of yellow fever virus with that of other viruses, physician Thomas Monath asserted that yellow fever virus arose as a unique virus 3,000 years ago. If this is true, then references to hemorrhagic fevers before 1000 B.C.E. must be evidence of diseases other than yellow fever. Yellow fever virus infects both monkeys and humans. In humans fever, chills, muscle aches, and headaches—a quartet of symptoms for many infections—mark the onset of yellow fever. These symptoms may last days before subsiding. The fortunate recover, whereas in others the symptoms recur. Yellow fever virus attacks the kidneys and liver, and the victims develop jaundice. Internal bleeding marks the final stage of yellow fever, and then the victims die.

Yellow fever originated in West Africa and through the slave trade came to the Caribbean and the Yucatán in the 1640s. From these points of origin, trade brought yellow fever as far north as Philadelphia in 1793. The South, with its long, warm summers, was especially vulnerable to the disease.

For centuries people were at a loss for the cause of yellow fever. Some blamed foul air, others sewage, and still others the all-embracing term *filth*. In 1881 Cuban physician Carlos Findlay posited that mosquitoes transmit yellow fever. The loss of more American soldiers to yellow fever than to battle wounds in the Spanish American War led the U.S. Army to seek the cause of yellow fever. Working with the conjecture that filth caused yellow fever, the army began in Havana in 1899 a campaign of hauling away garbage, sweeping the streets, and cleaning up yards. The absence of yellow fever in the early months of 1899 seemed to confirm their hypothesis, but by December yellow fever had returned to Havana, infecting 1,400 people.

In response to this outbreak, the army established a yellow fever commission headed by physician and major Walter Reed, who revived Findlay's mosquito hypothesis and shrewdly focused efforts on *Aedes aegypti*. Reed allowed the mosquitoes to bite volunteers, two of whom died of yellow fever. Having proven the mosquito a vector for yellow fever, he aimed to eradicate it from Havana. Fumigating homes with the gases of ignited sulfur and ignited tobacco stems, the commission killed large numbers of mosquitoes. Perhaps more important was the effort to prevent *Aedes aegypti* from reproducing. Knowing that the mosquito laid her eggs in clean standing water, Reed ordered the receptacles of water that served for drinking and bathing to be covered. Reed's subordinates drained other standing water or put kerosene in it. So successful were these measures that the last case of yellow fever in Havana was in September 1901.

To the extent that the people of Louisiana paid attention to Walter Reed's work, many were skeptical. They had seen hypotheses about what caused yellow fever go in and out of fashion. Reed's was just the latest in a long line of claims about yellow jack, a slang term for the disease. The members of New Orleans's city board of health were more receptive to Reed's work. Its members proposed that local volunteers identify the places where mosquitoes bred as a first step toward eradicating them. Nearly every house had a cistern, which caught and stored rainwater for drinking and bathing, making enormous the task of eliminating the mosquitoes' breeding spots.

Recognizing that the cisterns were ideal as a breeding ground for mosquitoes, Dr. Quitman Kohnke of the health board recommended that the board pour kerosene in the cisterns. The oil would float on the surface of the water and would contaminate only the thin film of water at the surface. Because spigots drained the cisterns from the bottom, they would still make clean water available for drinking and bathing. The film of kerosene would deter the yellow fever mosquito from laying her eggs in the cisterns and would drown mosquito larvae by preventing them from breathing air at the surface. Kohnke also urged the board to cover all cisterns with a fine wire mesh to prevent *Aedes aegypti* from laying her eggs in them. These

Walter Reed secured his fame in 1900 in the campaign against yellow fever. He hoped to become U.S. Surgeon General, but his death from appendicitis in 1902 truncated his career. Courtesy: Library of Congress.

two steps—adding kerosene to the water and covering the cisterns with a wire mesh—would be central to the effort against yellow fever.

In August 1901 the health board chose one neighborhood to test Kohnke's strategy. The effort was intended to gauge public support for an antimosquito campaign rather than to reduce the mosquito population. With only a single neighborhood in compliance, the mosquito would have plenty of other places to breed. The effort went badly from the start. With no ordinance to compel compliance, all the health board could do was cajole. Many homeowners refused to participate, and several of those who did expressed dissatisfaction, claiming that the water treated with kerosene tasted and smelled of the oil. The botched campaign convinced Kohnke that the health board would get nowhere without a law compelling residents to kerosene and screen. But with yellow fever in abeyance, the city council refused to pass such a law.

Between 1900 and 1904, New Orleans recorded no deaths from yellow jack. In 1905, however, the disease flared in Panama and Belize, and because New Orleans imported fruit from Central America, the city was in danger of being affected. Acting tardily, the city suspended imports from Central America in May 1905. The origin of yellow fever in New Orleans in 1905 is difficult to pinpoint; it is likely, however, that a shipment of fruit from Central America that spring harbored a viruliferous mosquito, which bit one of the workmen unloading the fruit. Doubtless, the first cases aroused no suspicion, since the symptoms of yellow fever are easily confused with those of other ailments.

The workmen were Italian immigrants who knew nothing about yellow jack, who were not fluent in English, who distrusted outsiders, and who lived in close-knit communities that resisted penetration from outsiders. The Italians' deep suspicion of doctors and hospitals were based on rumors circulating in their native land that stated that whenever cholera spread through the population, the Italian hospitals euthanized the patients who were too ill to have survived the disease. No evidence supports the claim of euthanasia, but these Italian immigrants, strangers in New Orleans, apparently retained prejudices that stopped them from alerting a physician at the first case of hemorrhagic fever.

Only in July, perhaps two or three months after it had reached New Orleans, did physicians first learn of cases of yellow fever. On July 12 an Italian family called a physician to attend to two members of the family with yellow fever. The two died, and the physician, alarmed by what he saw, reported to the health board that he had witnessed two "suspicious" deaths.[3] The family refused the health board's request for an autopsy, but the secrecy had been lifted. On July 14 a physician examined two more corpses and identified in them the symptoms of yellow fever. Yet he hesitated, deciding that the evidence was inconclusive. Members of the health board knew better. Although they made no announcement of yellow fever, they fumigated every house within several blocks of the tenements in which the victims had died. From Walter Reed's work the authorities in New Orleans had learned that the fumes from ignited sulfur were toxic to *Aedes aegypti*. Two more deaths on July 18 led the health board to alert health departments in Mississippi, Alabama, and Texas of a disease "presenting symptoms of yellow fever."[4] By July 22 New Orleans had registered one hundred cases of yellow fever and twenty deaths, prompting the health board at last to announce that the city was in the midst of a yellow fever outbreak. The next day the newspapers in New Orleans carried an article by Kohnke and Dr. Joseph White of the U.S. Public Health and Marine Hospital Service. The two urged that residents kerosene and screen, empty all containers of stagnant water, oil cesspools, sleep under mosquito nets, and screen doors and windows.

The people of Louisiana reacted to yellow fever in a variety of ways. The residents in New Orleans' 14th Ward, apparently in the belief that filth caused yellow jack, cleaned streets, yards, and gutters. Parents sang their children a yellow fever lullaby.

Little Gold Locks has gone to bed,
Kisses are given and prayers are said,
Mamma says, as she turns out the light,
Mosquitoes won't bit my child to-night.[5]

Into August the Italians continued to resist the efforts of physicians to treat the sick. Whenever someone died under the care of a physician, the family accused him of murder. Blacks, denied the opportunity to join with whites in fighting yellow fever, formed their own organizations. The residents of Vinton, Louisiana, refused to accept their mail, believing it to be contaminated with yellow fever. The U.S. Postal Service responded by closing the post office in Vinton, thereby suspending mail delivery. Some railroads stopped carrying mail to Louisiana.

On July 26 the health board dispatched one hundred volunteers in groups of five to find people with yellow fever, fumigate their homes, and kerosene and screen. That day, New Orleans's mayor hired one hundred men to clean and flush city gutters. On August 1 the city council passed an ordinance requiring residents to kerosene and screen within 48 hours. Those who refused to comply risked a $25 fine or 30 days in jail or both. That the city jailed some residents reveals that well into the outbreak, acceptance of the mosquito hypothesis was not universal. In mid-August Dr. Felix Fermento addressed an Italian audience in Italian, and his work, along with that of Catholic clerics, began to lessen the distrust the Italians felt toward the medical profession.

Medical authorities expended great effort educating the residents of New Orleans about yellow jack and the steps they could take to lessen their vulnerability to it. Nearly every evening, physicians delivered public lectures. In October they turned their attention to the schools, hoping the children would take the antimosquito message to their parents. The newspapers urged readers to kerosene and screen. Clergy preached sermons against yellow fever, exhorting their members to combat *Aedes aegypti*. Activists wore lapel pins that stated: "My cisterns are all right. How are yours?"[6]

Yellow fever swelled in the sultry months of July and August, peaking on August 12 with more than one hundred cases reported that day. Thereafter, the rate of infection and the number of deaths gradually declined. By October the number of new cases of yellow fever had dropped so low that President Theodore Roosevelt felt it safe to visit New Orleans on October 25. Physicians reported the last case in November. In total, physicians reported 9,321 cases of yellow fever and 988 deaths. In 1906 Louisiana had only one case of yellow fever. In 1909 New Orleans completed its water system. The cisterns no longer necessary, the city council outlawed them. Science had vanquished yellow fever.

DISEASE AND NUTRITION

Novelist and Nobel laureate William Faulkner peopled his novels with characters that were lazy, dull, and venial, a stereotype that was more than fiction thanks to hookworm and pellagra, which plagued Southerners in the early 20th century. Science conquered hookworm first and in doing so changed the lives of Southerners. In 1902 U.S. Department of Agriculture zoologist Charles Wardell Stiles traced "lazy man's disease" to a worm that lived in the soil and burrowed into the feet of a person who walked barefoot on infested ground.[7] The worm entered the veins, which carried it to the lungs. An infested person coughed up the tiny worms in his sputum and, in swallowing the sputum, inadvertently brought them to his intestines, where they once more borrowed into soft tissue, feeding on blood and multiplying into the thousands. In this way hookworm caused anemia, listlessness, and retardation. An infested person excreted worms in his feces, and since in the South many outhouses were nothing more than the seat of a toilet above exposed ground, hookworm spread from feces through the ground to infect still more people.

Stiles took a no-nonsense approach to hookworm, understanding that Southerners could easily prevent its spread by wearing shoes and building sanitary outhouses. At a conference sponsored by President Theodore Roosevelt's Commission on Country Life, Stiles in 1908 approached Wallace Buttrick, secretary of the Rockefeller Foundation's General Education Board, for help. Agreeing with Stiles that Southerners could prevent hookworm with a few simple changes in their lives, Buttrick in turn approached John D. Rockefeller, who in 1909 established the Rockefeller Sanitary Commission with a $1 million budget over five years. The commission met resistance at first. Some southerners resented a northern patrician's interference in their lives. Others took offense to the suggestion that southerners had worms. Still others thought the commission used the specter of hookworm to scare southerners into buying shoes, which the commission intended to sell, thereby profiting from the misfortune of the South. Yet dissenters did not carry the day. Atlanta, Georgia, alone raised $3 million in 1910 to launch its own campaign against hookworm. Newspapers ran articles urging rural southerners to wear shoes and to construct sanitary outhouses. Public schools sent home leaflets explaining what parents could do to prevent hookworm in their children. The commission set up demonstration stations throughout the South where people could look at hookworm larvae under a microscope. These efforts paid dividends as southerners changed their daily habits by wearing shoes and constructing self-contained outhouses whose contents a township periodically evacuated, disinfected, and disposed of.

Pellagra was less easy than hookworm to combat. From pellagra's first appearance in the South in 1902, medical science would need more than thirty years to vanquish the disease and thereby change the lives of Americans. Pellagra manifested several symptoms that are easy to confuse with the symptoms of other diseases. The red rash of pellagra mimics sunburn, and the diarrhea that was sometimes the only symptom of pellagra in a mild case mimicked innumerable intestinal disorders. In severe cases pellagra produced psychosis and thoughts of suicide, symptoms that suggested mental illness more readily than pellagra. No less mysterious was the cause of pellagra. Its prevalence among the poor led early investigators to focus on the lifestyle of the poor. Several researchers commented on the monotony of the poor's diet, the ubiquity of corn in the diet led scientists to finger it as the culprit. Studies of cornmeal isolated fungi and mold and seemed to strengthen the argument that consumption of corn, or at least moldy corn, caused pellagra. Southern boosters of the agrarian way of life asserted that if their citizens were to eat corn, they should be sure to consume fresh, home-grown corn. Some advocates of this position pointed to corn grown out of state, particularly in the West, as the culprit, an accusation that revealed more about regional animosities than about disease. A second group of scientists expected pellagra to be yet another disease caused by a pathogen. These were the scientists who advocated the germ theory of disease and who saw in the work of bacteriologists Robert Koch and Louis Pasteur a method of identifying and combating the pathogens that caused disease. But the efforts of these scientists to identify the putative pathogen that caused pellagra made little headway. The incidence of pellagra in insane asylums and nursing homes only superficially seemed to suggest that pellagra was a transmissible disease. Yet the nurses in these places were in daily contact with pellagrins but never became ill, undermining the suspicion that pellagra was an infectious disease.

The road to the cause and cure of pellagra began not in an asylum or a silo of corn but in a laboratory of the Wisconsin Agricultural Experiment Station, where in 1914 nutritionist Elmer V. McCollum was feeding rats to determine with precision the elements of a healthy diet. McCollum noted that rats fed butter remained healthy, whereas those on margarine ailed. He reasoned that butter must contain an element not found in margarine and which is essential to health. Isolating a compound from butter, McCollum christened the nutrient Vitamin A. McCollum's discovery launched the search for more vitamins and created the modern science of nutrition. Among the discovery of other vitamins was that of nicotinic acid (niacin) in 1937. Niacin cured pellagra, sometimes bringing about improvement only hours after pellgrins ingested it. Thanks to this discovery, the number of deaths from pellagra in the United States dropped from 2,040 in 1940 to 865 in 1945.

The success of niacin against pellagra led to calls for vitamin fortification of food. World War II amplified these calls as Americans recognized the value of a healthy citizenry to fight the war abroad and to staff the factories at home. The partisans of nutrition touted the slogan "vitamins will win the war," and Abbott Labs sold "victory vitamins."[8] At the suggestion of nutritionists, McCollum among them, millers added vitamins to flour, pasta, and rice, and dairymen irradiated milk to increase its content of Vitamin D. Eager to tout the wholesomeness of breakfast cereals, their manufacturers likewise added vitamins to cereals, though critics continued to lament their high sugar content.

The success of vitamins in curing pellagra, rickets, and other diseases of vitamin deficiency spurred Americans to supplement their diet with vitamin pills in their quest for good health. The vitamin pill was part of the daily regime of the middle class and college educated more than of the working class and high school graduate, suggesting to one historian that middle-class Americans were more proactive than their working-class counterparts in managing their health. This zeal led to paternalism. One factory owner gave his workers vitamin pills during the winter to reduce absenteeism. As this example shows, Americans have, since the discovery of vitamins, believed with varying degrees of credulity that vitamins prevent the common cold. Particularly seductive in this context was the belief that if a small amount of vitamins is good, more must be better. An advocate of this logic, Nobel laureate in chemistry and peace Linus Pauling announced in 1970 that megadoses of Vitamin C might prevent colds. Responding to this announcement, that winter Americans bought Vitamin C tablets faster than druggists could stock them. Americans likewise have taken large doses of Vitamin E in the belief it may prevent angina and heart attack. Worried that their children might not be getting all the vitamins they needed in their diet, mothers insisted their children take a multivitamin tablet as part of their morning routine. Athletes took vitamins in the quest for superior performance. Eager to lose those last five pounds, dieters, most of them women, took vitamins because they helped the body metabolize fat. Other Americans took vitamins in the quest for sexual potency, for more energy at work, for greater concentration, and as a protection against stress and irritability. So widespread was the appeal of vitamins in the 20th century that Americans bought vitamin-enriched soaps and lotions.

SCIENCE, SMOKING, AND THE FITNESS CRAZE

In its battle against cigarette smoking, medical science confronted an addiction in the lives of millions of Americans. Advertisers and popular culture glamorized smoking as an activity of virile men and seductive women. In the 20th century cigarette smoking permeated all strata of America, from street sweepers to celebrities. Americans smoked for a variety of reasons and often started young. Some succumbed to advertising, others to peer pressure, and still others took up smoking as a nervous habit. Many Americans found in smoking a pleasurable way of relaxing after dinner or on break at work. Once hooked, people found the habit of smoking difficult to break, because the body comes to crave the nicotine in tobacco. As early as the 1920s, some lone physicians and scientists condemned smoking, but generally medical science was slow to mobilize against it. Only pressure from the American Cancer Society, the American Heart Association, the National Tuberculosis Association, and the American Public Health Association forced U.S. Surgeon General Luther Terry to act. In 1962 he formed a commission to study the link between smoking and cancer, and in 1964 its report, "Smoking and Health," was unequivocal in fingering tobacco as a carcinogen. In coming to this conclusion, medical science offered a definition of causation that differed from what bacteriologists and virologists meant by causation. The germ theory of disease offered a straightforward definition of causation: the presence of a pathogen causes disease in an infected body. Medical science could not, however, apply this simple definition of causation to smoking. Not only did a pathogen not cause cancer in smokers, making moot the claims of the theory, but not every smoker developed cancer. Rather, smokers had a higher rate of some cancers—lung cancer prominent among them—and of heart disease than nonsmokers and so were more likely than nonsmokers to develop these diseases. In the case of smoking, medical science came to understand causation as probability rather than certainty.

In the short term, tobacco companies claimed uncertainty about the hazards of smoking and had little trouble critiquing a probabilistic definition of causation. By the mid-1970s, however, medical science's attack on smoking began to affect the habits of Americans. Between 1975 and 1985, smoking fell by a greater magnitude than in any decade before or since. In the 1980s states and localities began to ban smoking from public places, restricting the geography of smoking. As with vitamins, the science against smoking found greater adherents among the middle class than among the working class. By 1983 less than one-third of middle-class men and women smoked, compared with 43 percent of working-class men and 38 percent of working-class women. Acting against what they perceived as the norms of middle-class culture, working-class Americans continued to smoke in large numbers. Although science could shape the lives of Americans by decreasing the number of smokers, it could not extinguish smoking from American culture in the 20th century.

In the quest for good health, science did more than condemn smoking. It encouraged people to be active. In 1968 cardiologist Kenneth Cooper published *Aerobics* and in 1972 *Aerobics for Women*, urging Americans to exercise. Himself a jogger, Cooper recommended the activity for its simplicity and ease of undertaking. Jogging required no equipment, training, or facilities; one simply plodded along the road at a comfortable pace.

Americans took to jogging during their lunch break, forgoing a caloric meal with their colleagues. Joggers spoke of exercise as a type of meditation, a way of relieving stress in an America that they believed fostered anxiety in the daily routine of work and suburban life. Hardcore enthusiasts set up jogging dates to court prospective partners and entered 10-kilometer races on the weekend.

Science intersected with economic reality as gasoline prices rose in the mid-1970s. The physicians who had extolled the virtues of jogging now applauded Americans who bicycled to work as a way of saving gasoline and boosting fitness. Ebbing in the late 1970s, the fitness movement rebounded in the 1980s as mountain biking became popular in the West and adolescents joined their high school soccer team. Yet the fitness movement was never as widespread as its proponents wanted to believe. Like the crusades for vitamins and against smoking, the fitness craze, a middle-class movement from the start, never penetrated far into the working class. Whatever benefits science might ascribe to exercise, many Americans believed they were too busy to lace up their shoes for a daily jog.

CHILDBIRTH AND CHILD CARE

Science affected the lives of women and men by shaping the experience of childbirth in the 20th century. Central to the experience of women was the movement of childbirth from the home to hospital, the new center of medical science and technology. In 1900 only 5 percent of births were in a hospital, but by 1950 the percentage had risen to 88. As early as 1906, American physicians began using a German method of inducing semiconsciousness, the twilight sleep, in women giving birth. Under the tenets of this method, physicians drugged women with morphine and scopolamine and reduced auditory and visual stimuli by keeping women in darkened rooms and away from noise. Women who underwent twilight sleep could not recall the pain or the experience of childbirth; they had surrendered these sensations to the science of pharmacology. The twilight sleep movement gained adherents in the early 1910s as popular magazines reported it to millions of readers. The death of Mrs. Francis Carmody, an advocate of twilight sleep, during childbirth in 1915 weakened the movement but not women's reliance on drugs to ease the pain of childbirth.

Women took narcotics and barbiturates to relieve pain even as the movement toward natural childbirth gained adherents in the late 1940s. Since the 1970s women have had, in addition to these drugs, epidurals to block the sensation of pain. For some women an epidural changed their role in childbirth from participant to spectator, prompting the analogy that if giving birth to a child is like pulling a rabbit out of a hat, the epidural has relegated the mother's role to that of the hat rather than the magician. Other women reported dissatisfaction at having succumbed to the pain of childbirth by having an epidural rather than by enduring the pain to give birth naturally. Some nurses coaxed women to have an epidural by comparing natural childbirth to having a tooth drilled without anesthetic, and the American Society of Anesthesiologists added to this chorus in 1995 by issuing a report that women who had an epidural eased the anxiety of their husbands. The science of pharmacology and the technology of an epidural not only managed the pain of childbirth, they branded it as undesirable for the mother, husband, and hospital staff.

More than ease pain, science in the 20th century could dictate the onset of labor. Oxytocin and similar drugs allow women to induce labor. Some women choose induction as a convenience to end the discomfort of the last days of pregnancy or to accommodate the schedule of an obstetrician who had planned to leave town near the due date. In some instances oxytocin merges convenience and economy in allowing women with a due date around January 1 to induce labor at the end of December to qualify for the tax deduction of having a dependent.

Like science, medical technology has shaped the experience of childbirth. The sonogram allows parents to learn the sex of their child before birth and to pick a name and decorate the nursery accordingly. The electrical fetal monitor records not only the heartbeat of the fetus by the onset and duration of the mother's contractions, tempting husbands and nurses to watch the machine rather than the mother for signs of progress. Some women, feeling like an extension of a machine rather than a person, find the experience of childbirth alienating under these circumstances.

Having given birth to her baby, a woman has the task of feeding it. The science of nutrition, changing how Americans ate in the 20th century, also shaped what women fed their infants. Women who wanted a substitute for breast milk turned to vitamin-fortified formula. Some women found in formula a convenience, others comfort in the putative nutritional superiority of formula over breast milk, and still others turned to formula because they had problems breast feeding their infant or because they had to return to work.

By deemphasizing the value of intuition and advice from relatives and friends, science shaped the way women raised children. In a phenomenon one historian termed "scientific motherhood," (Apple) mothers in the 20th century increasingly turned for advice on childrearing to doctors and popular literature from the medical community. Beginning early in the century, women's magazines published advice columns written by physicians. Contemporary magazines continued this trend at the end of the century. In addition to these articles, parents-to-be read child-care books. Especially popular was physician Benjamin Spock's *Baby and Child Care* which had by 1985 sold more than 30 million copies.

Since the beginning of the 20th century, physicians and scientists, too, have sought to educate mothers. As early as 1910 physician Josephine Baker, alarmed at what she perceived as substandard mothering among the immigrants and poor in New York City, founded Little Mothers' Clubs in an attempt to teach the latest science on raising children to the young girls who were often responsible for the care of younger siblings while their mother worked. By 1912, 20,000 girls in New York City attended club meetings, and in the 1920s Wisconsin offered a similar program, teaching young girls to mix formula and to bathe infants. In the 1920s and 1930s, public schools embraced the notion of teaching young girls the science of caring for children by establishing courses in home economics. During the later 20th century gender lines blurred. Some boys took home economics and more girls took woodshop and auto mechanics.

Medicine Grew in Prestige and Cost

In 1900 life expectancy in the United States for men was 46.3 years, and for women 48.3 years. In 1999 these figures surged to 73.9 years for men and 79.4 years for women. Medicine, along with improvements in sanitation, deserved much of the credit for this increase in life expectancy. Vaccines and antibiotics cut the death rate for infectious diseases. Childhood killers retreated with the advance of pharmacology. Medicine, secure in its successes, grew in the 20th century to command government funding and the research budgets of universities. Medicine also grew in wealth and prestige. Young men and women aspired to become physicians and thereby to enjoy a large income and high status. The expansion of medicine in the 20th century led to opportunities for medical students to specialize. So pronounced did the trend toward specialization become that some specialists regarded themselves as superior to the general practitioner.

As medicine grew in influence, it routinely made news. Print, radio, and television journalism summarized the results of medical studies. Americans paid particular attention to those studies that recommended inclusion of a food or beverage in the diet or touted the value of exercise. In some cases controversy resulted. One study, in announcing the value of exposure to sunlight, ran the risk of confusing Americans who for years had avoided sunlight or used sunscreens because they feared exposure to ultraviolet light. In its capacity to make news, medicine had the power to shape public opinion. With the medicinal establishment against the use of tobacco, the public turned against cigarette smoking—once a glamorous act but by the late 20th century a vice.

Medicine grew not only in influence but in cost, often exceeding the ability of Americans to pay in full. In the 20th century insurance companies arose to pay the cost of medical care, but in an effort to maximize profits, insurance companies labored to curb the cost of medical procedures. Despite the attempt at containment, medical costs rose faster than the rate of inflation in the 20th century, creating an unsatisfactory situation. The physician who as part of a heath maintenance organization agreed to fixed fees for the procedures he performed reduced the time he spent with each patient, to maximize the number of patients he saw and the procedures he performed. Patients, feeling shortchanged, were unhappy that their physician spent so little time with them. Medicine in the 20th century was not merely a science and a profession. It became a business.

NOTES

1. Paul Starr, *The Social Transformation of American Medicine* (New York: Basic Books, 1982), 191.

2. Allan M. Brandt, *No Magic Bullet: A Social History of Venereal Disease in the United States Since 1880* (New York: Oxford University Press, 1985), 172.

3. Jo Ann Carrigan, *The Saffron Scourge: A History of Yellow Fever in Louisiana, 1796–1905* (Lafayette: University of Southwest Louisiana, 1999), 170.

4. Ibid.

5. Benjamin H. Trask, *Fearful Ravages: Yellow Fever in New Orleans, 1796–1905* (Lafayette: University of Louisiana at Lafayette, 2005), 105.

6. Ibid., 114.

7. Elizabeth W. Etheridge, *The Butterfly Caste: A Social History of Pellagra in the South* (Westport, Conn.: Greenwood Press, 1972), 210; Rima D. Apple, *Vitamania: Vitamins in American Culture* (New Brunswick, N.J.: Rutgers University Press, 1996), 13.

8. Ibid., 10.

4

THE CITY

Cities are ancient, predating even the invention of writing. From their origin in the Near East, cities spread throughout Eurasia and Africa. Cities arose independently in the New World but were not numerous before the arrival of Europeans, who, settling the eastern seaboard, founded Boston, New York City, Philadelphia, and Charleston. From this modest beginning, cities spread west in the 19th century as Americans migrated across the continent. In the 20th century cities in the Sunbelt gained population at the expense of cities in the Rustbelt.

Cities accommodated their burgeoning population by growing vertically as well as horizontally. The skyscraper was a source of civic pride, an architectural as well as a technological achievement. As cities grew in space and in population, the need for transportation became acute. The automobile was one solution, but it congested streets. The problem with the automobile was that each car moved only a small number of people, often just one, from place to place. Better to carry large numbers of people to their destination: so arose mass transit in the form of the subway, the electric railway, and the bus. These were as much manifestations of technology as they were forms of transportation.

The settling of large numbers of people in cities led to the spread of disease. Plague descended upon San Francisco in 1900 and yellow fever upon New Orleans in 1905. Cholera outbreaks were common in many cities. Public health officials responded by improving sanitation, notably by chlorinating drinking water and disposing of trash. Cities became places to live and work, not cesspools in which to die.

THE CITY IN 1900

A nation of farms at its inception, the United States in 1900 had cities with millions of inhabitants along the East and West coasts and the Great Lakes. In 1900 40 percent of Americans lived in cities of at least 2,500 people, and 20 percent lived in cities with more than 100,000 residents. Nearly 6.5 million people (10 percent of the population) lived in the three largest cities: New York City, Chicago, and Philadelphia.

Aggregates of people, metal, concrete, and brick, America's largest cities accommodated their growing populations by thrusting their buildings into the sky. Before the invention of the elevator, buildings could not be much more than seven or eight stories, the number of floors a person would be willing to climb. The invention of the elevator in the mid-19th century overcame the limit of human fatigue, and buildings crept skyward. Functioning as a skeleton, the steel girder made possible heights that architects could scarcely have envisioned in the 19th century. Strengthening the wall, the steel girder allowed builders to use large windows to let in sunlight. As a rule, no point within a room was more than thirty feet from a window, the distance to which sunlight penetrated a room. Lights, mounted to the ceiling, provided additional light, though the incandescent lamp stationed on the desk provided too little light to be worth the trouble. In 1939 builders began to supply offices with fluorescent lights because they had the advantage over the carbon or tungsten filament light of keeping cool. The building boom of the 1920s culminated in the Empire State Building in 1931. Buildings of this size upset the daily life of people who lived downtown. Very tall buildings sent rent in an upward spiral, pricing people who lived downtown out of their apartments. Unable to afford rent, a city dweller who lived downtown could either lower his or her standard of living by moving into a tenement or maintain the same standard of living by moving farther from downtown. The city thereby grew horizontally as it grew vertically.

By 1900 science and technology had had little effect in making comfortable and sanitary the lives of millions of urbanites. Jane Addams, founder of Hull House, dispensed with the pleasantries one might have expected from a middle-class woman in describing the conditions of daily life in Chicago: "The streets are inexpressibly dirty, sanitary legislation unenforced, the street lighting bad, the paving miserable and altogether lacking in the alleys and smaller streets, and the stables foul beyond description." New York City fared no better. Rain transformed the roads, clogged with garbage, into "veritable mud rivers," commented one journalist.

The traditional method of attacking filth did not enlist the help of science and technology but instead labeled it a sign of immorality. This line of reasoning blamed people who lived amid squalor for their own misery. Because many of the poor lived in unsanitary conditions, their critics had an easy time correlating poverty and unsanitary habits. Responsible for their actions, the argument went, the poor had chosen to live amid filth rather than in cleanliness.

In 1900 the alternative to sermonizing was the germ theory of disease, which held that filth, particularly organic waste, was a breeding ground for the bacteria that caused disease. Because insects that fed on this garbage could transmit disease to people throughout the city, no one was safe in this world of microbial invaders. The surge in the number of deaths from typhoid in July and August 1902 in Chicago brought the city near panic. In August alone 117 people died of typhoid fever and more than 1,000 had been infected. Health officials and doctors, knowing a waterborne bacterium caused typhoid, fingered the water supply as the culprit and wondered how this microbial tragedy had happened given that engineers had reversed the direction of the Chicago River to carry sewage away from Lake Michigan. Amid hysteria, the schools shut off the drinking fountains, for fear that they might spread the disease, and required students to carry their own water.

The drop in deaths from typhoid that fall did not end the problems urbanites faced in their daily lives in the first years of the 20th century. In an era before the automobile was widespread, innumerable horses provided transportation in the city. Worked long days with little regard for their health, horses dropped dead in the thousands. As late as 1912 10,000 horses died in the streets of Chicago. Scavenger crews were to remove dead horses before decomposition progressed far, though in some cases the horses were left in the streets several hours, more than enough time to make them carrion for the flies that made miserable the lives of rich and poor alike, since living apart from the crowded downtown was no protection against flies and the microbes they carried. Likewise full of pathogens were the 600,000 tons of manure that horses deposited every year on the streets of Chicago, according to the Automobile Chamber of Commerce. So much manure strained the capacity of American cities to remove it. Horse carts rode over the manure that was not removed and that baked in the sun, grinding it into dust that polluted the air, entered the lungs, and clung to surfaces, both inside and out.

The poor and immigrants had little defense against disease in the cities, often living in overcrowded tenement houses. In 1900 Manhattan had more than 42,000 tenements and 1.6 million dwellers. New York City's Eleventh Ward housed nearly 1,000 people per acre, a density that may have been the highest in the world. Several families often crowded into a single room, sharing one toilet and breathing unventilated air. Near the factories where they worked, tenement dwellers breathed a high concentration of the smoke generated by coal-burning factories. Visitors to Pittsburgh and Chicago complained that the smoke screened out sunlight, creating a nightmarish atmosphere.

In 1900 the technology of road building accommodated the horse. City streets had wooden blocks, often cedar, covered with gravel. Granite block was more durable but also more expensive than wood. The Chicago Department of Public Works criticized granite block for wearing down horse hooves and vehicle wheels. Whether wood or granite blocks, engineers held that horseshoes caught the crevices between the blocks, improving traction. Asphalt was an alternative to wood or stone block. Bicyclists, numerous in the cities in the first years of the 20th century, preferred the smoothness of asphalt to the unevenness of blocks and urged cities to pour asphalt over the streets' old surfaces. Health officials joined the campaign for asphalt, noting that its smooth surface would be easily swept clean of manure, dirt, and other debris. By 1902 American cities, responsive to these pleas, had paved 16 percent of their streets with asphalt. By 1909 the figure had risen to nearly one-third. The spread of asphalt and the increase in the volume of traffic in the cities moved children out of the streets where they had played and onto sidewalks and porches.

THE TECHNOLOGY OF PUBLIC TRANSPORTATION AND DAILY LIFE

An alternative to the horse was the electric railway, though in 1900 it had not replaced the horse but instead competed with it for traffic. The result was a tangle of pedestrians, electric railways, and horses in the downtown of large cities that writhed with traffic. In 1900 the streetcar was ubiquitous as a means of public transit. In 1902 New York City had 400 miles of track and 400 million passengers a year, Chicago had 700 miles of track and 300 million passengers a year, and Philadelphia's streetcars carried 325 million passengers a year. In addition to the trolley, New York City and Chicago had elevated trains: New York City with more than 100 miles of track and 200 million passengers a year and Chicago with half the number of passengers. Overall, the number of people who used public transit more than doubled between 1900 and 1910. In the early years of the 20th century, Americans traveled long distances by train, and in 1910 1,400 trains used the north and south terminals in Boston every day, 514 used the Broad Street Terminal in Philadelphia, and 322 used Union Station in St. Louis.

The obvious solution to the congestion aboveground was to build a subway below ground. Although it was not the oldest, the New York City subway, in operation for all but four years of the 20th century, came to symbolize all that Americans liked and disliked about their subways. In taking Americans to work, the theater, a baseball game, or a restaurant, the subway shaped daily life in the city.

Running seven days a week, twenty-four hours a day, the New York City subway had a profound effect on the daily lives of millions of New Yorkers. The subway opened to fanfare, with New York City mayor George B. McClellan, the namesake and son of a Civil War general, delivering a ceremonial speech on October 27, 1904. After the speech, he was to have switched on the lead car, then given the controls to a motorman. Instead, full of adrenaline and having a good time, McClelland took the controls, guiding the subway for five miles. New Yorkers took to their subway as readily as had their mayor. Between 2 P.M. and 6 P.M. that first day, 15,000 people—representatives of the city's upper class—rode the subway, relinquishing it after 6 P.M. to the masses. From then until midnight, 150,000 New Yorkers thronged the subway. In its first full year of operation the subway carried 106 million passengers.

New Yorkers, celebrating their subway's effects on their daily lives, danced the Subway Express Two Step and sang "Down in the Subway." With music in their ears and on their lips, New Yorkers need not have ridden the subway to feel its magnetic pull. "Doing the subway" became the phrase for riding for fun rather than with a destination in mind. For these New Yorkers, the subway was not about getting to work but about meeting people and visiting places in a kind of tourism offering Jack, no longer a dull boy, new possibilities for recreation. In the 1920s more than 1 million New Yorkers a year took the subway to Coney Island. The subway also had lines to Yankee Stadium, Ebbets Field, and the Polo Grounds. The subway cars became the place where partisans assessed the virtues

Pedestrians and automobiles share an intersection in Detroit. In many cities pedestrians challenged cars for control of the road, jaywalking and, when they had numbers in their favor, stopping traffic while they crossed the road. Courtesy: Library of Congress.

and faults of their teams. Several baseball players, Brooklyn Dodgers shortstop Pee Wee Reese among them, even took the subway to and from Yankee Stadium.

Popular as the subway was, from the outset the fear circulated that being underground, the people who rode the subway did not have access to fresh air. Accounts of dizziness, nausea, and fainting reinforced this fear, as did an article in the *New York Medical Journal* that air in the subway harbored viruses and the bacteria that caused tuberculosis and pneumonia. The people who used the germ theory of disease to amplify the threat to public health of horse manure and of horse carcasses in America's urban streets now turned a critical eye to the unventilated air in the subway. City Hall fought back, hiring a Columbia University professor to gauge the subway's air quality and churning out leaflets entitled "Subway Air as Pure as Your Own Home."

A second complaint raised the charge of sexual harassment. When a subway car was full, riders without a seat crowded together in very tight quarters. The movement of a subway car imparted a rocking motion to the crowd that mimicked the motion of the hips during sex. Women who found themselves beside a strange man during these episodes thought themselves violated and sought redress. The Women's Municipal League of New York urged the city in 1909 to reserve one subway car for women. The city declined, but a separate line, the Hudson and Manhattan Railroad, which operated a subway between New York City and New Jersey, agreed, reserving for women one car on each train during the peak hours between 7:30 A.M. and 9 A.M. and between 4:30 P.M. and 7 P.M. The experiment lasted only two months, at the end of which the Hudson and Manhattan Railroad bowed to criticism by men, and even some women, that the practice of exclusion was inherently unfair.

This episode made clear that the subway shaped daily life in bringing together rather than separating people of diverse backgrounds. The subway was its own melting pot as immigrants rode with native-born Americans, and the rich with the working class. The daily lives of New Yorkers intersected on the subway, if nowhere else.

The number of riders, having risen steadily until 1930, dipped during the Great Depression and rebounded during World War II, when gasoline rationing coaxed New Yorkers who owned a car to take the subway. But despite fluctuations in the number of riders, the subway was so entrenched in the daily life of New Yorkers that two-thirds of the city's residents in 1945 did not own a car. During the postwar boom, however, even the subway found it difficult to compete with the automobile. The number of riders declined with the increase in car ownership, and by the 1970s the New York City subway tottered on the brink of anarchy. Gangs roamed the subway, and idle youth gave voice to their frustration in graffiti. The city spent $800,000 in 1971 to remove graffiti only to need $1.3 million the next year. In 1979 a group of volunteers, the Guardian Angels, formed to patrol subway cars against gangs. New art replaced the graffiti as New Yorkers, their daily lives enmeshed in the rhythm of the subway, began talking of expanding it.

THE CITY ACCOMMODATES THE AUTOMOBILE

The New York City subway and mass transit in general lost riders after World War II as Americans took to the roads in the automobile. The reader may recall that chapter 2 introduced the automobile as a technology of transportation. Here we will consider the relationship between the automobile, the city, and daily life. A luxury in the first decade of the 20th century rather than a mass commodity, the automobile nonetheless added to the volume of traffic, the very problem the subway was introduced to alleviate. Cities responded to the automobile by seeking to impart order to an ever-increasing volume of traffic. In 1903, with the automobile only 5 percent of traffic, New York City issued its first traffic laws, requiring drivers to signal a turn with their hand. Slow traffic was to occupy the farthest lane to the right, with faster cars to the left. Cars were to park along the curb and in the direction of traffic. They were not, however, to park within 10 feet of an intersection. Many drivers thought these laws violated their freedom and ignored them. Cities responded by deputizing police to enforce these laws. In 1904, the year New York City opened its subway, Philadelphia formed its first traffic police. By 1920 New York City had some 6,000 such officers. In 1905 New York City painted lines on its roads to guide traffic. In 1911 Detroit erected the first stop sign at an intersection, and the next year Salt Lake City posted the first electric traffic light, operated by an officer who threw a switch. In 1922 Houston, dispensing with manual operation, posted the first automatic traffic light. By 1929 New York City had some 3,000 automatic traffic lights.

In their quest for mobility, Americans gravitated to the automobile. In Chicago, for example, the ratio of cars to residents rose from one automobile for every sixty-one residents in 1915 to one car for every eight residents in 1930. That year in Seattle and Detroit, the ratio was one to four, and in Los Angeles, one to three. In 1931 435,000 people in Los Angeles used the car for work, and only 262,000 used public transit.

The popularity of the automobile increased the problem of congestion in the city. At first drivers parked their cars at the curb alongside the store or office they wished to visit. But curb parking narrowed streets, exacerbating the problem of traffic jams. In the 1920s cities began to outlaw curb parking and instead to build parking garages for the storage of cars while their owners were at work or shopping. Chicago's Pure Oil Garage opened in 1926. It had twenty-two stories of parking with three elevators to ferry cars up and down the levels. In 1925 Detroit had three parking garages with the capacity to store 1,500 cars at 75 cents a day each. Larger than their counterparts in Detroit, the parking garages in Cleveland could store

7,000 cars. As the banning of curbside parking chased drivers from street to parking garage, retailers, conscious of the need for parking, moved to the outskirts of the city, seeking more land.

By then, Los Angeles had more cars per person than any other large city and anticipated the national trend by favoring cars over mass transit. To accommodate the sixfold increase in the number of cars in the 1930s, Los Angeles built new roads and, after 1939, a freeway system. The first city to define itself by the automobile, Los Angeles was also the first city to suffer from smog, a combination of *smoke* and *fog,* a term coined by the London *Daily Graphic* in 1947. Smog is the detritus of the automobile, which emits, among several hundred chemicals, nitrogen oxide and hydrocarbons. Sunlight converts these chemicals into ozone, a part of the upper atmosphere, but in the lower atmosphere they form smog. Other cities that teemed with cars also found themselves suffocating in smog. New York City issued smog warnings to alert residents with respiratory problems to stay indoors, California set limits on auto emissions in 1966, and Los Angeles and Washington, D.C., set aside dedicated freeway lanes for buses in hopes of encouraging their use.

Futurists, aware that cities were up to their skyscrapers in cars, yearned for the day when Americans could work from the home rather than commute to the city. In the late 20th century, workers in information technology, able to do their job as productively at home as at the office so long as they had a computer, began to work at home. Freed from the commute to the city and the structure of the office, these workers were able to arrange their daily lives as they desired. In other circumstances, the line blurred between work and leisure, making it difficult for some to get away from work. No longer could telecommuters go home to get away from work—they were already home. Others, aware of their supervisors' expectations, put in long hours to boost productivity. Still others, feeling detached from their colleagues, commuted to the city at least one day a week.

Telecommuting did not end the problem of congestion and the need for parking space for all the cars that came into America's cities. In the 1980s Dallas had the highest ratio of parking space to office space in the United States. Thanks to this ratio, the average resident of Dallas parked just two and a half blocks from work. Had they been willing to walk farther, the parking space would have been available for other uses.

THE AUTOMOBILE, PEDESTRIANS, AND THE CITY

Aside from occasional accidents pedestrians in 1900 had little trouble contending with the slow horse and carriage. As the century wore on, the automobile replaced the horse and carriage, and pedestrians found that this new technology limited their mobility. They had to wait at the traffic lights at intersections. City planners timed the lights to give cars more time with a green light and pedestrians less time to cross streets. The technology of the traffic signal required pedestrians to walk at a brisk pace, between three and five miles an hour, just to cross a street before the signal changed. Pedestrians in America's largest cities reacted by ignoring the traffic signal and paying attention instead to the flow of traffic. When an opening between cars appeared, pedestrians dashed across an intersection in violation of the traffic signal.

Sometimes pedestrians challenged cars for control of the road, and in some places pedestrians enjoyed an advantage over the automobile. The roads of Boston's financial district have twists and turns that slow down cars and embolden pedestrians to challenge them by crossing a street en masse. By sheer numbers, pedestrians were able to stop cars. Boston's pedestrians signaled cars with their hands when they wished to cross a street. Cars that did not stop caused pedestrians to slap their hoods in a display of anger. By one estimate, 60 percent of Bostonians crossed a street against a traffic signal. The situation was similar at the tip of Manhattan, whose streets the Dutch had laid out in the 17th century. Dark and narrow, these streets deterred cars and encouraged pedestrians to congregate along them. Elsewhere in New York City, as in Boston, pedestrians, when numbers were in their favor, brought cars to a halt by brazenly crossing a street. The tussle between pedestrian and automobile sometimes ended tragically for the pedestrian: flesh was no defense against steel. New York City averaged 7 pedestrian fatalities per 100,000 residents, Chicago 4.5 per 100,000, and Los Angeles 4 per 100,000. Of those killed, 68 percent were men.

The automobile was not the only technology against which pedestrians pitted their will. In the late 20th century, Americans in large cities put the bicycle to a new use in ferrying documents from office to office. These bicycle messengers, paid by the delivery, had an incentive to travel as rapidly as possible. By averaging thirty to thirty-five miles per hour, bicycle messengers could earn $250 to $300 per day in the mid-1980s. They could only average such speeds by chasing pedestrians from the streets, yelling at them and blowing whistles to bully them aside. The tendency of bicycle messengers to go whichever way they pleased on a one-way street forced pedestrians to look both ways down a one-way street before crossing. Even with the greatest caution, accidents occurred. In 1986 three people died and 1,640 were injured in accidents between bicycle and pedestrian.

With so many automobiles and pedestrians contesting the same space, congestion was inevitable. Here again, city planners favored the automobile, awarding it more space then pedestrians had. Between 1915 and 1930, Chicago widened and built 112 miles of new streets. In the 1920s Pittsburgh spent $20 million to widen its streets. New York City's Lexington Avenue was fifty feet wide, whereas sidewalks were only twelve and a half feet wide. Worse, street vendors, trash containers, and mailboxes reduced Lexington Avenue's sidewalks at some points to just six or seven feet in width.

To relieve pedestrian congestion, cities built underground concourses and overhead skyways. At least one sociologist has argued that city planners designed these structures to benefit the automobile, which then had fewer pedestrians to compete for access to the roads. In Dallas pedestrians dined at the salad bars and pizza places along the concourse. In New York City the Rockefeller Center concourse, and the skyways in Minneapolis and St. Paul, protected pedestrians against winter weather. In Minneapolis four skyways fed into the IDS Crystal Court and were as congested as the streets. In Chicago the University of Illinois Circle Campus skyway won few pedestrians because it was uncovered and so offered no weather protection. In Charlotte, by contrast, pedestrians used the skyway more than the streets. Concourse and skyway appealed to young professionals, whereas minorities and the poor remained on the streets.

SCIENCE, TECHNOLOGY, WATER, AND DAILY LIFE

As have people at all times and in all cultures, Americans needed water for their daily activities of drinking and bathing, and they needed it on a vast scale. In 1900 Chicago's waterworks pumped 500 million gallons of water a day, enough for every resident to have 139 gallons. That their water ought to be clean was obvious, but how was one to define clean? People judged water by its taste, odor, and turbidity, and these judgments varied by person. There was then no objective standard of water quality. This state of affairs changed as public health officials adopted the germ theory of disease: to be clean, water had, in addition to the above qualities, to be free from pathogens.

American cities first sought to purify water of pathogens by filtering it. Before 1900, the Massachusetts Board of Health's Lawrence Experiment Station had demonstrated that it could purify water by filtering it through soil or sand. But because soil was useful for so many other purposes, cities in the early 20th century adopted a sand filter for their water. The incidence of typhoid decreased in cities that used a sand filter, demonstrating sand's ability to trap pathogens. The second method was chlorination. It was well-known in the 19th century that the addition of chlorine to water killed all microbes in it. Because chlorine was a poison, American cities moved cautiously in adopting it. In 1908 Chicago chlorinated its sewage before discharge but did not take the obvious step of chlorinating its drinking water. The next year, Jersey City, New Jersey, became the first American city to chlorinate its drinking water. As with filtration, the number of typhoid cases fell with the chlorination of water. The success of filtration and chlorination sparked a public health movement in which public officials encouraged Americans to drink more water, in the belief that, within reason, the more water a person drank the healthier he or she would be. These officials hardly needed to add that the more water a person used to bathe, the cleaner he or she would be.

Before 1900, Americans paid a flat rate for water and could use as much of it as they pleased with impunity. By 1900, however, the water meter, installed in many homes, recorded how much water people used and so enabled the city to charge for water by volume. Public health officials lauded the water meter for preventing people from wasting water, but many Americans resented the meter as yet another intrusion on their daily routine and as an attempt to charge as much as possible for water. The Free Water Association of Portland, Oregon, noting that air was free, retorted in 1907 that water should be free because it was necessary for health and hygiene. With the population of cities increasing in the 20th century, however, public health officials and sanitary engineers could see little alternative to metering water, since wastefulness might deplete water to dangerous levels. By 1920 60 percent of American cities metered their water.

The conviction grew in the 20th century that water should not only be free from pathogens, it should contain chemical elements that benefited health. In 1923 Rochester, New York, was the first city to add the halogen iodine to water in the form of the salt sodium iodide to prevent goiter. In 1945 Grand Rapids, Michigan; Newburgh, New York; and Evanston, Illinois were the first cities to add the halogen fluorine to water in the form of the salt sodium fluoride to prevent tooth decay. Some citizens' groups were skeptical of fluorine's benefit, noting that sodium fluoride was a rat and insect poison and wondering whether it was toxic to humans. Anecdotes surfaced that fluorine produced white spots on enamel, "Texas teeth," that turned brown over time. During the cold war, opponents of fluoridation branded it an assault on personal freedom and blamed communist sympathizers for promoting it. As controversy intensified, sixty cities stopped adding sodium fluoride to their water between 1953 and 1963, though twenty-six of these cities renewed the practice after 1963. A new round of charges and countercharges plagued the issue in the 1990s. The American Dental Association and other professional organizations endorsed fluoridation and pointed to reductions in tooth decay in cities that had added sodium fluoride to the water. Opponents, however, attempted to link fluoridation of the water supply with aging and even with AIDS.

The campaign against fluoridation, sordid as it was, nonetheless revealed that urbanites were scrutinizing their water supply, or at least what the media told them about it. In 1991 New York City assemblyman John Ravitz pointed to reports of pollutants in the water, most sensationally the presence of bacterium *Escherichia coli*. Amid a health alert, New Yorkers turned to buying bottled water to meet their need for drinking water. They were not alone. Sales of bottled water in the United States tripled in the 1990s, and the marketing and sale of bottled water gained a momentum of their own as health-conscious Americans were eager to believe that bottled water was superior to tap water, much as the typhoid scare of 1902 had led children to carry bottle water to school. Even in cities that could document that their water had fewer impurities than bottled water, sales of bottled water nonetheless increased, thanks to skillful marketing.

The provision of water to urbanites was only half the job, though. Cities also had to remove vast quantities of waste. The practice of dumping sewage into lakes and rivers was expedient and inexpensive but defied both common sense and germ theory. Because sewage added pathogens to the water, the dumping of it into bodies of water could only spread disease. By 1900 no one could doubt the link between raw sewage and the incidence and spread of typhoid. One attempt at a solution was to dilute sewage with water before discharging it into lakes and rivers. Better was Chicago's decision in 1908 to chlorinate sewage to kill pathogens before discharging it into lakes or rivers. More ambitious was the city's reversal of the Chicago River, and the sewage it carried, away from the waterworks. Paper was an easier matter than sewage, since American cities burned trash. Organic waste had no simple remedy. Some health officials, seeing the solution to an urban problem in the countryside, advocated the feeding of organic waste to pigs. This idea gained adherents during World War I when the federal government urged farmers to produce more food to feed both civilians at home and soldiers abroad. Agricultural scientists retorted, however, that pigs fed waste produced poorer cuts of pork and ham. and were more vulnerable to cholera than pigs fed on corn. These scientists doubted that the food supply would long remain safe if the feeding of organic waste became widespread among ranchers. World War I at an end, cities moved toward landfills as the depository of the tons of waste they generated every day.

In Cleveland, Ohio, residents long complained about the filth and stench of the city, but the complaints had little effect on sanitation. Businessmen, averse to paying for technology to clean up their environs, resisted legislation that would have required them to clean up what they spewed into the air and water. Matters came to a head in 1969, when the Cuyahoga River caught fire. Industry had turned the river into a cesspool of flammable waste, which spontaneously ignited. The Environmental Protection Agency ordered businesses that discharged waste into the river to double their capacity to treat sewage. Having resisted for decades spending money on cleaning what they spewed into the environment, businesses at last had to spend millions of dollars on waste treatment plants. The victory for Cleveland's residents in cleaner air and water was real but hollow as industry and jobs left the city for the Sunbelt and Mexico. Bereft of employment Cleveland deteriorated into a vortex of crime, drugs, and apathy. The city had become a study in pathology.

EUGENICS AND THE CITY

In 1883 British mathematician Francis Galton coined the term *eugenics* from the Greek *eugenes* for "wellborn." Whereas his cousin Charles Darwin had identified natural selection as the mechanism of evolution, Galton believed humans could transcend natural selection by shaping their own evolution through selective breeding. The mating of the intelligent and industrious would yield the wellborn that Galton prized, but the mating of degenerates would produce criminals and the insane. The goal of eugenics was to encourage the breeding of superior humans and to discourage the breeding of inferior people. At its extreme, eugenics barred some groups of people from marrying one another and from having children, thereby shaping in dramatic fashion the lives of ordinary people.

To the extent that eugenics was a science, it was an offshoot of genetics. The founder of genetics, Gregor Mendel, showed that parents passed genes to offspring. Genes code for traits. For example, Mendel demonstrated that a single gene codes for the color green in peas. A single gene also codes for the color yellow in peas. The geneticists who followed Mendel had no difficulty extrapolating his findings to the rest of life. Of particular interest was the role of heredity in humans. In a casual way, people had long appreciated the importance of heredity, noting for example that a child looked strikingly like his or her mother. Geneticists sought to formalize observations of this kind, tracing, for example, the transmission of the gene for brown eyes through several generations of a family. In the course of this work it was natural for geneticists to wonder whether intelligence and traits of character were inherited with the lawlike regularity that Mendel had observed with simple traits in peas.

A tradition going back to John Locke viewed the intellect and character of people as unformed at birth. The experiences of life craft the traits that a person develops in interaction with the environment. American psychologist B. F. Skinner took this view to its logical extension, claiming that given a child, he could raise him to be a criminal or a physician or anything between these two by manipulating the environment.

Eugenicists claimed the opposite, believing that heredity rather than the environment shaped people. The children of criminals grew up to be criminals not by modeling the behavior of their parents but because they possessed the genes of criminality. Eugenicists were as sure that genes existed for criminality or laziness or prodigality and for lawfulness or diligence or thrift as they were that genes existed for eye color. Just as with eye color, eugenicists supposed that a small number of genes, sometimes only a single gene, coded for intelligence or character traits. Heredity was destiny, and no amount of education or moral upbringing could steer a person with the genes for criminality away from a life of crime. In such instances the education of criminals wasted resources that would better go to educating people with the genes for intelligence and other desirable traits. The world for the eugenicist was as simple as it was black and white. Characteristics were not nurtured over the course of one's life. They were set rigidly and inescapably at conception. Over the course of a lifetime, a person simply realized his unalterable genetic potential.

After 1900 eugenics spread to the United States. In 1909, with funding from the Carnegie Institute, former University of Chicago professor Charles Davenport established the Eugenics Record Office to advance the study of the relationship between heredity and mental traits, intelligence foremost among them. Davenport hired as his assistant Harry H. Laughlin, who had been teaching agricultural science at Kirksville Normal School in Missouri. By the 1920s Laughlin would be powerful enough to influence Congress as well as public opinion.

As did other eugenicists, Laughlin looked with suspicion at the cities. He saw in them the prisons and asylums that housed what he believed to be degenerates. Cities were also full of idiots, as the new intelligence tests, which purported to measure the mental age of people, made clear. Henry H. Goddard tested immigrants upon their arrival at Ellis Island. Many scored poorly, a circumstance Goddard attributed to low intelligence rather than to the fact that immigrants took the test upon arrival to the United States, a time when many of them were still adjusting to their new circumstances. Goddard declared feebleminded all who scored below the mental age of thirteen. Before long, eugenicists expanded the meaning of *feebleminded* to include all mental traits of which they disapproved.

Eugenicists looked with alarm at the number of immigrants who poured into America's cities. In 1900 immigrants were nearly one-third of the residents of New York City, Chicago, Boston, Cleveland, Minneapolis, and San Francisco. Between 1900 and 1910, 41 percent of the people who settled the cities were immigrants. By 1912 three-quarters of the population of Cleveland was first- or second-generation immigrant. Around 1910, the source of immigrants shifted from northwestern Europe to southern and eastern Europe. These were the new immigrants who performed poorly on the intelligence tests, and Goddard and Laughlin labeled them inferior to those from northwestern Europe. It is difficult to gauge the degree to which personal bias influenced Goddard and Laughlin's disapproval. The two men seem to have had a visceral dislike of immigrants, In any case, Laughlin quickly identified the immigrants from southern and eastern Europe as carriers of inferior genes. If the nation's leaders were not careful, these immigrants settling the cities, they claimed, would there breed with the native born, contaminating America's gene pool. Genetic contamination also came from a second source. Starkly racist in their attitudes, the most militant eugenicists asserted that African Americans, like the new immigrants, had inferior genes. These eugenicists abhorred the mixing of races in the cities and condemned miscegenation for contaminating the gene pool of whites.

Faced with these threats, eugenicists worked on three fronts. Laughlin sought to stop the immigration of people from southern and eastern Europe, and as Eugenics Expert Witness to the House Committee of Immigration and Naturalization in 1921, he shared with Congress the results of intelligence tests, which he believed proved the inferiority of these immigrants. During testimony, Laughlin had aides hold placards with photos of immigrants at

Ellis Island. These placards had captions that, without any evidence, labeled the immigrant inferior. These were the immigrants, Laughlin testified, who threatened America's gene pool if Congress failed to act. Congress passed the Immigration Act of 1924, reducing immigration from southern and eastern Europe to just 2 percent of their total population in 1890, a year in which few of these immigrants populated the United States.

On the second front, eugenicists sought to regulate which groups of Americans could marry one another. The selection of a mate is so basic a freedom that any law that restricted this right necessarily affected the lives of Americans in a dramatic way. Slave owners in the antebellum South had fornicated with their female slaves, but aside from these dalliances, custom was clear in separating blacks from whites. Eugenicists appealed to the aversion to blacks that many whites felt by sounding the alarm that blacks and whites might interbreed, especially in the cities where they were in close quarters, if left to their own devices. This interbreeding would be of the worst sort, for it would combine the inferior genes of blacks with the equally bad genes of low-class whites. The results of such unions were sure to be degenerates, it was claimed. Alarmed at this prospect, the state of Indiana in 1926 forbade the marriage of a white and a person who was at least one-eighth black. By 1940 nearly thirty states had antimiscegenation laws.

On the third front, eugenicists aimed to control reproduction rights. As with the choice of a marriage partner, the decision to have children is so basic that it affects the lives of all Americans. Eugenicists opposed the reproduction of the feebleminded, criminals, and the insane. Knowing that voluntary measures would go nowhere, eugenicists urged the states to sterilize mental and moral degenerates. Indiana passed the first sterilization law in 1907, and by 1931 thirty states had similar laws. Laughlin had helped draft some of these laws by writing in 1920 a model sterilization law that states could copy. A judge could order the sterilization of a person after conferring only with a physician and an attorney or family member representing the person. The sterilization laws, sweeping in their denial of a basic human right, did not go unchallenged. In Virginia there arose the case of three generations of a family whose female members had dropped out of school and had children out of wedlock. The sterilization of one of these women, Carrie Buck, led to the case *Buck v. Bell* (1927). Writing for the majority, Supreme Court justice Oliver Wendell Holmes upheld Virginia's sterilization law. Holmes wrote that "the principle that sustains compulsory vaccination is broad enough to cover cutting the fallopian tubes." With sterilization laws safe from further challenge, the states, with California in the lead, sterilized 64,000 Americans between 1910 and 1963. In many cases the person to be sterilized thought he or she was to have some routine procedure. Carrie Buck, for example, never understood why she could not have more children. Not only were these Americans deprived of their right to reproduce, they were denied the knowledge of what was happening to them.

Successful as it was in shaping the lives of Americans, eugenics had critics. The influential geneticist Thomas Hunt Morgan attacked eugenics in 1925. He thought the term *feebleminded* too broad and imprecise. Whatever it was, feeblemindedness, Morgan believed, had many causes, yet eugenicists treated it as though it had a single genetic cause. Here was the heart of Morgan's critique: eugenicists were wrong to ascribe to a single gene or at most a small number of genes the complexities of human thought and behavior. Complex thought and behavior arose in interaction with the environment, which might be either rich or poor in stimuli. The environment must therefore play a role in how people thought and behaved, but the single-gene explanation of eugenics was too simple to account for complexity. To the eugenicists' claim that heredity was destiny Morgan responded that both heredity and the environment are important in shaping humans. Morgan's commonsense position drew allies. Journalist Walter Lippmann doubted that intelligence tests measured innate intelligence. Morgan's former student and Nobel laureate Hermann J. Muller declared that only by equalizing the environment for everyone would it be possible to assess the degree to which heredity determined intelligence. These critiques might have come to naught as the musings of intellectuals but for the excesses of Nazism. The leaders of Nazi Germany adopted eugenics, much of it from the United States. Indeed, in 1936 Heidelberg University awarded Laughlin an honorary doctorate. Yet the Nazis went further than the United States in implementing the principles of eugenics. Between 1933 and 1937 Germany sterilized 400,000 inmates of prisons and asylums. Thereafter, German eugenics descended into barbarism as Germany euthanized citizens it deemed mentally and physically deficient. These policies shocked Americans and tainted eugenics, which never recovered in the United States the popularity it had enjoyed in the 1920s.

Technology Spreads out the City

Technology has challenged the identity of cities. Whereas people once lived in them because of the premium humans place on face-to-face exchanges of information, the telephone, fax machine and, above all else, the computer with Internet connection allowed Americans to substitute electronic communication for face-to-face interaction. In the process people have chosen to live and work outside the city. Rather than a dense core of people, cities have dispersed their population over a large area. White middle-class professionals left the city for the suburb. Retailers, conscious of the need for parking space, relocated to the suburbs, where they were within easy reach of customers. Cities that adapted to the new demographic and economic reality—Boston is an example—used technology to create jobs. Information technology and biotechnology may yet revitalize the city, pointing the way toward the future of work in urban America.

5

THE HOME

The traditional division of labor applied to American homes in 1900. Men worked outside the home and women worked, just as hard and in many cases harder than men, in the home. In lower-class homes women often worked outside the home as well, to add to the family income. In-the-home technology, over the course of the century, eased the toil of household chores. Electric-powered technologies transformed cooking and cleaning. Vacuum sweepers, electric irons, washers, and dryers were instantly popular.

Technology made the home a pleasant place in which to live. No longer were Americans resigned to being cold in winter and warm in summer. The furnace heated air by igniting natural gas, and electric-powered fans blew warm air throughout the house in winter. Electric-powered air conditioners cooled the home in summer.

Outside the home, the gasoline lawnmower cut grass to a uniform height. Herbicides killed crabgrass, dandelions, and other weeds. Science and technology made possible a monoculture of Kentucky bluegrass just as on the farm science and technology made possible a monoculture of corn, wheat, or another crop. Hybrid flowers brightened walkways and gardens. Comfortable inside and attractive inside and outside, the home was, as it had been for centuries, the focal point of American life.

ELECTRICITY AND DAILY LIFE

Electricity in the Home

Before widespread adoption of electricity, the home was dark and dirty by the standards of the 20th century. Light, produced by the burning of kerosene, wood, or coal left a residue of soot. In 1900 housewives lacked electric appliances to help them clean the home of soot and other dirt. Although men might haul water to the hand-cranked washer or take carpets outdoors to be beaten, the onus of housework, then as now, fell to women, who had to clean lamps of grime every few days, dust surfaces, wash floors, walls, and windows, and sweep carpets. Whether or not women hauled water to the washer, they scrubbed clothes by hand and ironed them under hot conditions, for the flatiron had to be heated on a stove and used on the spot. Should a housewife wish to iron clothes away from the heat of the stove she would find that the flatiron had lost heat in transit to a cooler place in the house and so was useless to iron clothes. Although women might have welcomed the heat from the stove in the winter, the heat discomforted them in summer. At three to twelve pounds, the flatiron taxed the endurance of women who had a pile of clothes to iron.

Electrifying the home was no simple matter. In the late 19th century, the debate between the proponents of direct current and those of alternating current complicated the delivery of electricity to the home. Thomas Edison led the campaign for direct current, a type of electricity produced by the rotation of a commutator in a magnetic field. The direct-current generators of the 19th century produced low voltage—120 to 240 volts—but these voltages did not translate into low prices by the standards of the early 20th century. Edison's customers paid 28 cents a kilowatt-hour in the 1880s. Edison built his first power station in 1882 and had 120 of them by 1887. Rapid as this expansion was, direct current had the drawback of limited range because of its low voltage. The electricity dissipated as heat, lowering the voltage, as it traveled through a wire, limiting the use of direct current to within a one-mile radius of a power station. The alternative to direct current was alternating current, a type of electricity produced by the rotation of a coil of wire in a magnetic field. Alternating current created higher voltages than did direct current. To maintain high voltage over long distances, engineer William Stanley invented in 1885 a transformer to boost voltage, and thereby counteract that which was lost through heat as electricity traveled through a wire. Near the home, a step-down transformer reduced voltage for use in the home. In 1893 George Westinghouse, a proponent of alternating current, won a contract to supply electricity to the World's Fair in Chicago. The 8,000 arc lights and 130,000 incandescent lights on display underscored that the United States had entered the era of electricity and signaled the triumph of alternating current over direct current. In 1895 a hydroelectric plant in Niagara Falls began supplying electricity in alternating current to homes and businesses twenty miles away in Buffalo, demonstrating the ability of alternating current to span distance. By 1900 alternating current had supplanted direct current in supplying electricity to the home.

The high cost of electricity made it a luxury in the first years of the 20th century, but Americans who could afford electricity were conspicuous in consuming it. The wife of Cornelius Vanderbilt, for example, greeted guests in a dress studded with tiny electric lights.

The Electric Light

The decrease in the cost of electricity from 28 cents a kilowatt-hour in the 1880s to 10 cents a kilowatt-hour in 1915 and to 7 cents a kilowatt-hour made what had been a luxury into a commodity Americans could afford. By 1927 more than 17 million homes had electricity, a number that comprised more than 60 percent of American homes and that equaled the number of electrified homes in the rest of the world combined. The spread of cheap electricity into homes across America transformed daily life. Electricity had its most immediate effect in lighting the home. Americans illuminated their homes with the incandescent light, which Edison had invented in 1879. Edison's light had a carbon filament in a vacuum, and the passage of a current through the filament caused it to glow. So durable was the carbon filament that the same lightbulb has lit Edison's home in Fort Myers, Florida, for 10 hours a day, 6 days a week, since 1927. General Electric engineer Irving Langmuir in 1913 developed a lightbulb with a tungsten filament in nitrogen gas. Cheaper to light than the carbon filament, the tungsten bulb supplanted Edison's light. In 1910 engineer and chemist Georges Claude invented the neon light, which glowed when a current passed through neon gas encased in a tube. In the 1920s several physicists realized that the passage of a current through a gaseous mixture of mercury and argon caused argon to radiate ultraviolet light, a type of light outside the visible spectrum. In the 1930s the addition of a fluorescent phosphor converted the ultraviolet light into visible light, and Americans in small numbers began to install fluorescent lights, which radiate 75 percent less heat, and therefore use less energy, than the incandescent bulb, in their homes.

The electric light freed Americans from the restrictions imposed by the length of the day, particularly during winter when the sun set early. The electric light allowed Americans to lengthen the day to suit their needs and, along with the clock (the first electric clock dates to 1918), to regulate their lives through technology rather than through nature. By illuminating the home at any hour of the day or night, the electric light stimulated children and adults to read, giving them knowledge and a perspective of the world wider than their neighborhood. The residents of Muncie, Indiana, for example, eager to read in the era of the electric light, went to their library, which increased its loans eightfold between 1890 and 1925. The spread of television after World War II and the Internet in the 1990s vied with books for the attention of Americans, though the electric light remained the source of illumination in the home. The monitor provided another source of light for Americans working on their computer. Even so, many people retained the habit of working in a lighted room, particularly

This woman empties the bag of her vacuum cleaner. Electric appliances like the vacuum cleaner reduced but did not eliminate the drudgery of housework. In many cases women put in long hours cleaning and cooking. Courtesy: Library of Congress.

when they tucked their computer away in the basement. So closely did Americans associate electricity and light that in the early 20th century they called power plants electric-light stations.

In replacing the burning of kerosene, wood, and coal for illumination, the electric light brightened the home. No longer did it need dark carpet to hide the residue of soot. Light-colored carpets, surfaces, and paint for the walls set homes in the 20th century apart from the dark interiors of the 19th-century home. To take advantage of light, architects designed homes with large open spaces in which one could pass from kitchen to dining room and to living room without going through a door. Each room had its own light affixed to the ceiling, sometimes as part of a fan, and a switch on the wall for turning a light on and off. The technology of indoor plumbing to bring water into the home and to remove waste allowed architects to consolidate toilet, sink, and bathtub in one room, whereas they had in the 19th century been separate rooms.

Electric Appliances

Technology did more than illuminate the home and shape its design. The spread of electric appliances lessened much of the drudgery of housework. Like the automobile, electric appliances spread rapidly throughout America—partly because electricity was cheap, but also because domestic servants were in short supply. Although the middle and upper classes routinely employed domestics in the 19th century, by 1900 the United States had only half the number of domestics as in Britain, and 90 percent of American families had no domestics. The movement away from domestic servants quickened in the first decades of the 20th century. Immigrant women took jobs as domestics when they could not find anything better, but the decrease in immigration during World War I and the surge in factory jobs meant that many immigrant women could trade their job in someone else's home for one in a factory. With domestics scarce, prices rose. By one estimate, a family in the 1920s needed $3,000 a year in income to afford domestic help at a time when the average family made only $1,000 a year.

The new appliances ran on an electric motor, the prototype of which Croatian American Nikola Tesla invented in 1899 by using two alternating currents to rotate a magnetic field and in turn a shaft that powered the motor. At the end of the 19th century, Tesla and Westinghouse, using Tesla's motor, invented the electric fan. A portable unit, the fan could be moved from room to room to circulate air during summer. Attached to the ceiling of a room, a fan served the same purpose; attached to a furnace, a fan could blow hot air through the ducts, bringing heat to the home in winter. By blowing air through the ducts, a fan circulated hot air quickly through the home, whereas before the use of a fan a furnace could circulate air only by heating it to 180 degrees Fahrenheit, a temperature high enough for the air to rise through the home, keeping everyone uncomfortably warm. The attachment of a fan to a furnace made the fireplace obsolete. Americans in the 20th century used the fireplace for show or built an artificial fireplace in the home. In this instance the fireplace went from being functional to being ornamental. Versatile, the electric fan spread throughout America despite its high price. A portable unit cost $5 in 1919, the price of a day's labor at Ford Motor Company. Once bought, however, a small fan was cheap to run at a 1/4 cent an hour.

Whereas the electric fan helped cool the home in summer and heat it in winter, the electric vacuum cleaner saved women the drudgery of hauling carpets outdoors to be beaten. Around 1901 James Spangler, a janitor at a department store in Ohio, added an electric fan to create suction from the mouth of a manual carpet sweeper. To catch the dust and dirt sucked up by the sweeper, Spangler stapled to a broomstick handle a pillowcase inside a soapbox, disposing of the pillowcase when it was full. Spangler's cousin showed the contraption to her husband, William Hoover, who began manufacturing the electric vacuum cleaner in 1907. In 1915 a small vacuum cleaner cost $30, and larger models $75. In 1917 demand was so robust that Montgomery Ward began advertising its models of vacuum cleaners in its catalog. By 1927 more than half of all homes with electricity had a vacuum cleaner.

The development of the nickel-chrome resistor in 1907 made possible a range of new electric appliances. The passage of a current through the resistor heated it. From this simple phenomenon arose the toaster, the coffee percolator, the hot plate, the hair curler, and the electric cooker, the incarnation of which in the late 20th century being the Foreman Grill. Perhaps the most immediate effect of the resistor on daily life was the replacement of the flatiron with the electric iron. At fewer than three pounds, the electric iron was easier to use than the flatiron. By 1927 more than three-quarters of homes with electricity had an electric iron. At $6 the electric iron was not inexpensive, but it was cheaper to use than heating the stove, a precondition to using the flatiron. By obviating the need to heat the stove, the electric iron kept the home cool in summer and freed the housewife to iron where she wished.

Given the close association between washing and ironing clothes, one should not be surprised that their invention as electric appliances dates from the same year, 1909. In 1910 Alva J. Fisher, an employee of the Hurley Machine Company of Chicago invented the Thor washing machine and in 1911 Maytag followed with its first model. Its first tubs wooden, Maytag converted in 1919 to an aluminum tub. Sales of electric washing machines rose from 900,00 in 1916 to 1.4 million in 1925. These early washers were not fully automatic, however. They had a hand-cranked wringer through which a housewife had to feed every piece of clothing. As late as 1948 Maytag's models still had a hand-cranked wringer, though in 1947 Bendix Corporation marketed a washer with a spin cycle that replaced the hand-cranked wringer for removing excess water. The Bendix model was the first fully automatic washer. One simply added clothes and detergent and, absent an electric dryer, hung them to dry. In the 1940s Americans began to use the clothes dryer to eliminate the chore of hanging clothes to dry and the dishwasher to eliminate the chore of washing dishes by hand.

The spread of electricity stoked the demand for shellac, an insulation to cover the copper wires that carried the current. Shellac—a deposit of beetles from only one part of the world, southern Asia—was scarce in the early 20th century. Because of this, Belgian American chemist Leo Baekeland determined to synthesize shellac in the laboratory. After three years and innumerable trials, Baekeland in 1907 combined formaldehyde with phenol to yield plastic, a material much more versatile than shellac. Durable and malleable plastic remade the home with everything from the exterior of microwave ovens to lawn furniture and from radios to alarm clocks.

POST–WORLD WAR II TECHNOLOGY

In 1919 Army officer Edwin George mounted a motor from a washing machine on a manual lawn mower. The spread of the gasoline-powered lawn mower quickened after World War II when Americans moved to the suburbs in large numbers. There the drone of lawn mowers announced the arrival of Saturday. By the 1960s the lawn mower was so ubiquitous that historian Charles Mills wrote in 1961, "If all of them [the lawn mowers] in a single neighborhood were started at once, the racket would be heard 'round the world."[1] The first mowers cut a 20-inch swath, though as they grew in power they grew in size, with the commercial models cutting 54 inches. In this sense the lawn mower did not merely cut grass, it created a business out of cutting grass.

An alternative to the gasoline lawnmower was the electric lawnmower, which was lighter and quieter and did not spew pollutants into the air. It might have won widespread use but for the fact that one had to run an extension cord from an outlet to the mower. The length of the cord restricted the electric mower to small yards, and maneuvering the mower so that it did not cut the cord taxed the patience of Americans who did not want the chore of cutting grass to be any more onerous than it already was.

Suburbanites who wanted a picture-book yard planted flowers that were the product of recurrent selection and treated their lawns with fertilizer and herbicides. Like a field of corn, the homeowner could use science to make his lawn a monoculture of Kentucky bluegrass. The quest for monoculture created a subtle, and sometimes not so subtle, pressure on homeowners. A dandelion in one yard, if allowed to seed, would infest other lawns, making imperative that that the homeowner with dandelions deluge his yard with herbicide. Chemlawn and True Green made their money on the fear homeowners harbored that their lawn, if left untreated, would not measure up to neighborhood standards.

Cooking was another chore that benefited from technology. The microwave oven owes its origin to World War II, when Britain used the magnetron to generate microwaves for a type of radar. Experimenting with microwaves, Raytheon scientist Percy Spencer serendipitously discovered that the waves melted a chocolate bar in his pocket. In 1947 Raytheon built its first microwave oven, though the company manufactured large units for restaurants, giving little thought to the home. In 1974 the Amana Refrigeration Company manufactured a home model, and by 1993 more than 70 percent of American homes had one. The microwave did not replace the stove and oven but filled its own niche. Americans preferred the taste of food cooked by traditional means and used the microwave to heat leftovers when they were in a hurry or a bag of popcorn when they wanted something to eat while watching television. The microwave was a technology of convenience that Americans used when they were too busy to gather around the table for a meal.

In the 1930s commercial food processors began freezing vegetables, then drying them in a vacuum until the ice crystals evaporated. The technology of freeze-drying yielded vegetables that mimicked fresh vegetables in their retention of flavor and nutrients, and the result spread throughout America's grocery stores in the 1950s. Freeze-dried vegetables were an alternative to canning, though in the countryside women still canned vegetables from the garden by hand. Suburban women preferred to buy frozen vegetables rather than undertake the chore of canning their own vegetables.

The smoke detector became a fixture in American homes after 1970. A 9-volt battery powers an ionization chamber and a horn. The ionization chamber emits a steady stream of ions, which the chamber measures as a current of electricity. When smoke enters the chamber it attaches to the ions, disrupting the flow of electrons. Sensing this disruption, the chamber sounds the horn, alerting a home's occupants to danger.

The compact disk (CD) improved the quality of music Americans played in their homes. The old phonograph records or audiotapes gave the listener both music and background noise. The CD, however, did not capture background noise because the recorder put the music into digital form. A laser reads the digitized music, encoded into the surface of a CD, as it passes along this surface. Because an instrument never touches a CD, it should, at least in theory, never wear out.

The cameras of the first half of the 20th century captured an image on film that had to be developed in a darkroom. In 1947 Polaroid Corporation founder Edwin Land invented the Polaroid camera, for the first time allowing the photographer to see a photograph within minutes of taking it. The digital camera took this trend toward contemporaneity to the extreme, allowing the photographer to see a photo the instant he took it. Parents could capture images of their children in spontaneous play, then download the images to computer and e-mail them to relatives in another state. In a sense the digital camera transformed the home into an impromptu photo studio, with sundry events the subject of photography.

SCIENTIFIC MANAGEMENT OF THE HOME

As much as the technologies of the 20th century helped the homemaker, they did not eliminate housework. Instead, advertisements and articles in magazines emphasized how appliances could help housewives raise their homes to a new level of cleanliness, thus creating higher expectations. These expectations drove women, according to three surveys in the 1920s, to spend more than fifty hours a week on chores in the home.

Women accepted these new expectations at the same time that they sought to use their time more efficiently. After visiting factories and offices, and having read mechanical engineer Frederick Taylor's ideas of scientific management, housewife and author Christine Frederick determined to apply the principles of scientific management to housework. Although Taylor's ideas were intended for the factory, Frederick chose the home as a type of domestic factory. Certainly it, too, was a place of production. Frederick made notes of her work habits and those of her friends. She times herself in domestic tasks and designed cabinets high enough so that she need not stoop to get pots and pans. She rearranged the utensils in her kitchen with an eye to efficiency. To spread the idea of efficiency in the home she wrote four articles in 1912 for *Ladies' Home Journal,* articles she expanded into her 1914 book, *The New Housekeeping: Efficiency Studies in Home Management.* Frederick also designed a correspondence course for women entitled Household Engineering. Other women wrote similar tracts in the name of household efficiency. Implicit in this work was a faith that quantification, a tool of science, was a means of improving efficiency. In a broader sense the application of scientific management to the home promoted a faith, typical of Americans, that science is a means of progress. The name Household Engineering implied that Frederick and other women sought to elevate housekeeping to an applied science with the home as their laboratory.

The idea that women could run the home on the principle of efficiency and by the methods of science led American colleges and universities to teach this new domestic science. Early in the 20th century, MIT and Columbia University began to teach home economics. But the movement toward scientific management of the home was confined neither to elite universities nor to the suburbs, where Frederick and her ilk lived. In 1925 Congress passed the Purnell Act, giving $20,000 the first year and incrementally increasing amounts through 1930 to the agricultural experiment stations to conduct "economic and sociological investigations" of "the rural home and rural life."[2] In the parlance of everyday speech, the Purnell Act funded the study of home economics and rural sociology. Congress's earmarking of this money for the agricultural experiment stations made explicit the connection between applied science and home economics, since the experiment stations had since the 19th century been thoroughly scientific and practical in their work. No less significant was the fact that the Purnell Act, in giving money to the experiment stations, targeted the rural home, bringing the science of domestic efficiency to the countryside. That the experiment stations were the research arm of the land-grant colleges led those colleges that were not already doing so to teach home economics, further expanding the discipline. Affirming the connection between applied science and home economics, housewife Mary Pattison founded a Housekeeping Experiment Station in Colonia, New Jersey.

COOLING THE HOME

In the 20th century Americans went from keeping food cool in an icebox to cooling food in an electric refrigerator. The principle of refrigeration dates from the 19th century. The kinetic theory of gases states that the molecules in a gas move at random, colliding with one another and with the container that holds them. Each collision releases heat, and it follows that a gas condensed by pressure will heat up because of the large number of collisions between molecules. The reverse is also true. A gas allowed to expand will cool because of the small number of collisions between molecules. This effect is apparent atop Mount Everest, where the low concentration of gases leads to few collisions between molecules and thus to very low temperatures. The electric refrigerator applied this effect to the cooling of food. An electric-powered compressor squeezed a gas into a liquid in a condenser, then pumped the liquid into a chamber where, in the absence of pressure, the liquid returned to a gas and expanded rapidly, cooling the chamber. Air blown across the chamber likewise cooled and, circulating in the refrigerator, cooled the food. The first refrigerators used ammonia as the gas to be compressed and expanded, though in the 1930s freon replaced ammonia.

In 1911 General Electric manufactured the Audiffren, its first refrigerator, and in 1926 the Monitor Top Unit, a larger refrigerator than its predecessor. In 1923 Seeger Refrigerator Company manufactured its first refrigerator, though not until the 1940s would it convert its refrigerator from wood to insulated metal. Crosley became in 1935 the first manufacturer to install in its refrigerators shelves for the placement of food. Sales of refrigerators leapt from 850,000 in 1930 to 4 million in 1941.

Air conditioning expanded the principle of refrigeration to the home. Rather than cool the small space of a refrigerator, an air-conditioning unit would cool the large expanse of a room or, in the case of whole-house units, of a home. In 1902 Willis Carrier, then an engineer at the Buffalo Forge Company, designed the first air conditioner to cool the room in which the company had its presses. In 1906 Carrier installed an air conditioner in a cotton mill in South Carolina. In 1924 Carrier installed air conditioning in the J. L. Hudson Company department store in Detroit, and in 1925 in the Rivoli Theater in New York City. Other sales to department stores, theaters, and hotels followed that decade as these businesses were quick to appreciate the value of air conditioning in attracting customers. In 1928 Carrier designed the first home air conditioning unit. That year he installed air conditioning in the House of Representatives, and the next year in the Senate.

Before air conditioning, the home had the electric fan to circulate air, but even this was little aid in sweltering summers. Air conditioning transformed the home into a place of comfort year-round and was particularly welcome in the South, where temperatures and humidity were high. Thanks to air conditioning, Americans could retreat to the home to escape the ravages of the summer sun.

SHOPPING FROM HOME

Television provided a forum through which Americans could shop from home. The commercial—standard fare on television since its inception—grew in the 1970s into the infomercial, a thirty-minute advertisement for a product. The more than tripling of cable television stations between 1972 and 1980 gave the airwaves lots of time to fill. Willing to sell time at discount, the stations were especially eager to fill the hours before prime time and the early-morning hours. Buying up time in the early morning, bodybuilder Jerry Wilson was the first to intuit the potential of the infomercial to coax Americans to spend money from home. Marketing an exercise machine as the Soloflex gym, Wilson's infomercial featured two actors, a man and woman, working out on the machine. The sexual tension between the two was meant to imply that the buyer would increase his sex appeal. The infomercial did not scare off buyers by advertising the machine's $2,000 price tag but instead announced the low monthly payment and the acceptance of payment by credit card. In the 1980s stations televised the Soloflex infomercial more than any other, and the infomercial's success gained it imitators. Several incarnations of abdominal machine promised rock-hard abs to their infomercial viewers. Merchants labored to disguise infomercials as programming so that channel surfers would not immediately click to another channel. By one estimate, an infomercial had two to ten second to catch viewers; otherwise they would change channels. Buyers placed more than three-quarters of orders within thirty minutes of an infomercial's broadcast. After a societal focus on health and nutrition led merchants to sell juicers by infomercial, the sale of juicers catapulted, in 1991, from $10 million to $380 million.

More than television, the computer and the Internet remade the experience of shopping from home in the late 20th century. In contrast to shopping at a bricks-and-mortar store, the Internet allowed Americans to shop anytime. They once had to visit several stores to compare products and prices, burning gasoline and time. The Internet made possible the comparison of hundreds, even thousands, of products and prices. At CNET, for example, one could compare prices, delivery dates, and product guarantees on computers and electronics. Shoppers could choose the best deal and purchase the product on the CNET site or print out the information to take to a local retailer in hopes of getting an even lower price. Large purchases were possible on the Internet. One could even buy cars and real estate online. At sites like match.com and eharmony.com one could even shop for a mate. Priceline.com eased the chore of comparing airfares by allowing a buyer to specify a price in hopes that an airline would be able to offer a seat at that price. Shoppers could sign up for e-mail notifications telling them about new products and sales from an electronic store. Web merchants could track the personal data and buying habits of shoppers so that an electronic store offered to sell a lawnmower to a suburbanite, Britney Spears CDs to teens, and a book on infant care to a new mother or father. A year later, that electronic store might offer a book on the care of toddlers to that same parent.

The mall, that secular Mecca of suburbia, had its counterpart on the Internet. The virtual mall, an aggregate of electronic stores, had several incarnations. In 1996 the Branch Mall had about 100 shops, and the Open Market had more than 8,000. The Branch Mall grouped vendors by category, such as food and drink, Virtumall listed its stores alphabetically, and iMall operated by keyword search. Typing in the word *wine,* for example, would generate a list of stores that sold various vintages of the beverage. Virtual malls provided a breadth of goods and services that one expected of a mall. The Branch Mall, for example, sold gourmet chocolates, contemporary Russian art, home security systems, and flowers, as well as offering services such as resume writing.

Narrower in its focus was the Electronic Newsstand, which sold magazines and newspapers. Americans could buy a kit for drafting a will from Global Network Navigator Marketplace. Electronic shoppers who wanted to buy Colombian coffee from Colombia had to pay in pesos. The conversion of dollars into pesos or more than fifty other currencies was possible at Global Network Navigator, which had a currency exchange link. A shopper at IONet Mall was able to receive a free life-insurance quote from its electronic store G. K. and Associate. The London Mall offered free quotes on medical, automobile, and homeowner's insurance. Engaged couples were able to register their wish list with the Ultimate Bridal Registry. At other electronic stores it was possible to buy a dog trained to assist the blind, a Victorian dress for a girl, a bridal gown, horror books and DVDs, Orthodox Christian icons, kits for brewing beer, fresh fruit from Florida and lobster from Maine, Halloween costumes, a pizza from Pizza Hut, groceries, a personalized astrological reading, the opinion of a psychologist or psychiatrist, in short, anything and everything.

Not all Americans shopped online. Women, who were three-quarters of buyers in the real world, were less than one-third of online buyers in 1996. Women perceived shopping as a social activity and felt uncomfortable in the solitude of shopping by computer. Women also liked to see and touch what they were buying, but the best the Internet could do was a photograph of an item for sale.

ENVIRONMENTAL CONCERNS

Yet the benefits of refrigeration and air conditioning came at a cost. In 1973 F. Sherwood Rowland and Mario Molina at the University of California, Irvine, reported that freon, the common coolant, damaged the ozone layer. Once in the atmosphere, freon breaks down under ultraviolet light to release chlorine. The chlorine in turn bonds with ozone, dropping it out of the ozone layer. Over time, freon has caused the ozone layer to thin. No less serious is the fact the freon may remain in the atmosphere for 150 years, meaning the freon emitted in 1990 may thin the ozone layer into the 22nd century. The thinning of the ozone layer is no academic problem, because as the ozone layer has thinned it has screened out less ultraviolet light, which is a carcinogen in large doses. Mindful of this danger, many people greeted with alarm Rowland and Molina's announcement. Manufacturers of refrigerators and air conditioners began in the 1970s to switch from freon to hydrochlorofluorocarbons, which break down in the lower atmosphere and so do not reach the ozone layer in large numbers. For every one hundred molecules of freon, only five molecules of HCFs will reach the ozone layer. In 1992 a group of utility companies offered $30 million for the design of an environmentally neutral coolant. In August 1993 Whirlpool claimed the prize with a derivate of HCFs that breaks down in the lower atmosphere even more readily than HCFs.

In the 20th century the burning of coal generated the electricity that powered the home. Only in the 1960s did scientists begin to gauge the harm of burning coal. When burned, coal releases sulfur dioxide and nitrous oxide, two chemicals that in the atmosphere combine with water vapor to produce acid rain. Action was needed, and under its clean coal technology program, the U.S. Department of Energy built an electric power plant in Brilliant, Ohio, that burned both coal and dolomite, a type of limestone. The dolomite absorbed sulfur before it bonded with oxygen, producing sulfur dioxide. Because the power plant is efficient at a lower temperature than a traditional coal-burning plant, it produces less nitrous oxide and carbon dioxide, a greenhouse gas. Another technology that reduced the amount of sulfur dioxide and nitrous oxide that escaped into the atmosphere was the spraying of power-plant smokestacks with a limestone mist. The limestone combined with sulfur dioxide to prevent its escaping into the atmosphere.

Along with the danger of acid rain, Americans in the 20th century came to understand that their electric-power plants, along with factories and automobiles, spewed carbon dioxide into the atmosphere. Plants use some of this carbon dioxide in photosynthesis, but the rest rises into the atmosphere, where carbon dioxide traps sunlight by preventing its passing into space after having reflected from earth. The rise in global temperatures, either a normal fluctuation or the result of global warming due to the buildup of greenhouse gases, threatens to melt the glaciers, thereby raising the ocean and flooding coastal cities. Electricity is not without perils.

Everyone in the Home Did Not Benefit Equally from Technology

The technologies that lessened the drudgery of housework were a mixed blessing. To be sure, they lightened the housewife's load, but in many cases housewives found themselves, even with the new technologies, working longer hours than their husbands. With the new technologies came higher standards of cleanliness that goaded women to clean more thoroughly then had their mothers and grandmothers. Some women found fulfillment in the daily routine of cooking and cleaning, but others, as Charlotte Perkins Gilman and Betty Friedan made clear, found the routine of household chores stultifying. The brunt of housework still fell heavily on women who worked outside the home, and they found themselves saddled with working two shifts, one at work and the second at home. The man of the house perhaps felt he had contributed his quota of labor by caring for the lawn. Technology did not benefit everyone in the home equally.

NOTES

1. Quoted in Time Books. *Great Inventions: Geniuses and Gizmos; Innovation in Our Time* (New York: Time Books, 2003), 124.

2. 7 U.S. Code, Section 361A.

6

COMMUNICATION AND MEDIA

We communicate with one another through a variety of technologies. Face-to-face is the oldest and perhaps the most satisfying form of communication, but in the 20th century technology enabled communication in other ways. The telephone, a 19th-century technology, permitted instantaneous communication with an immediacy matched only by face-to-face interaction. In the 20th century technology expanded the range of the telephone by making it portable. The result was the cell phone, whose use expanded exponentially in the late 20th century. The century witnessed the instantaneous transmission of the human voice not merely by telephone but also by radio. Television pioneered the instantaneous transmission of images along with sound. E-mail, growing faster than any other form of communication in the late 20th century, combined the intimacy of the letter with the speed of transmission across space. The Internet gave Americans unprecedented access to information, and its chat rooms brought people together in the electronic version of face-to-face communication. Taken together, these technologies made communication richer and more varied in the 20th century than it had been at any time in the past.

CINEMA

The use of a camera to take photographs in rapid succession and their projection onto a screen created cinema, a technology that transformed communication and entertainment and, with them, the lives of Americans in the 20th century. In 1888 Thomas Edison and his assistant Thomas

Dickson invented the kinetoscope motion picture camera. Others invented cameras that could likewise take a sequence of photographs without the photographer having to manually take each picture. Rather than having the inventors of motion picture cameras competing against each other, the Motion Picture Patents Company pooled the patents of motion picture cameras. The company collected a fee from each film and distributed a percentage to the patent holders. By 1920, however, independent film producers in Hollywood, California, had eclipsed the company.

Before television, cinema gave Americans a glimpse of a world larger than their neighborhood. Americans in San Francisco, Chicago, and Boston were able to watch the same movie, and in this way cinema created a uniform popular culture and a uniform message about how celebrities lived onscreen and off. Americans read in magazines the behavior and lifestyle of movie stars and emulated the behavior they saw on-screen.

In 1900 Americans watched movies in nickelodeons, a name that coupled the price of admission with the Greek word for theater. Concentrated in the large cities, the first nickelodeons had not in 1900 spread to the smaller towns that dotted the United States. The first movies were silent and short. Viewers watched, for example, people dancing or a train speeding along its route. These movies lacked a narrative or didactic purpose. They simply allowed Americans a diversion from the workaday world. Between movies, Americans watched live acts. The first theaters were a mix of silent movies and boisterous vaudeville.

In 1903 *The Great Train Robbery,* an 11-minute movie, created a new type of cinema. The rapid pace, the gunfight, and a narrative structure left a lasting imprint on cinema. One did not need to know English to follow the action of *The Great Train Robbery* and its ilk. One did not even need to know how to read. Cinema, in making few demands of its viewers, attracted immigrants and the working class, the bulwark of the audience in the early years of motion pictures.

By the 1920s the demographics had shifted toward Americans with more education and earnings than the typical immigrant had. Only one-quarter of the population had a high school diploma or a college degree, but this segment now made up the majority of theatergoers. Increasingly drawn from the middle class, the typical moviegoer was under age 35. Young and affluent, the Americans who went to the theater mirrored, however remotely, the youth and wealth of movie stars.

As Americans came to identify with movie stars, they began to emulate their behavior. Theatergoers learned the rituals of courtship, even how to kiss, from watching the virile men and seductive women who starred in the movies. As a result of cinema, some Americans became intimate earlier in their relationships than they might otherwise have. Americans thronged the theater to see Rudolph Valentino and Greta Garbo, whose sexuality titillated a generation of Americans. A Swede, Garbo seemed to personify the sexual freedom Europeans enjoyed in contrast to the conservatism of American sexual mores.

And cinema went beyond flouting conventional mores. Actors were featured drinking alcohol during Prohibition, and one study revealed that the hero of a movie was more likely than the villain to consume alcohol, lending respectability to the defiance of Prohibition. In the on-screen version of America, drinking alcohol was widespread and without stigma. Americans who adopted the values in these movies could scarcely find fault with drinking despite the Constitutional ban of it.

In 1927 Warner Brothers released *The Jazz Singer,* the first Hollywood feature length movie with sound. Its popularity ensured the permanence of talkies in American culture and made vaudeville obsolete. Whereas viewers had felt free to talk during silent films, creating a sense of community, movies with sound demanded silence from the audience so that listeners would not miss dialogue. Ushers now hushed members of the audience who persisted in talking. The result according to one historian was a diminished sense of community in the theater.

In the 1930s Disney created an alternative to traditional cinema with its animated films. Mickey Mouse, Snow White and Cinderella peopled the fantasy world of Disney, which many Americans, troubled by the events of the Great Depression, craved. Cinema, as the examples from Disney suggest, became escapist during the Depression. *Gone with the Wind* and the *Wizard of Oz,* both released in 1939, owed their popularity to their success in creating worlds that never existed.

After World War II, television added to the allure of cinema by giving Hollywood a medium through which to advertise its films. Advertisers packed commercials with action scenes from their latest movie, leaving viewers eager to see the entire film. Whereas the first theaters had been in the cities, the postwar boom led to the construction of theaters in the suburbs. Taking in a movie became part of the routine of suburban life. Teenagers, having gotten permission from their parents to take the family car, went to the movies on a weekend night. Boys and girls paired up on dates, and some couples were proud of the fact that they spent more time kissing than watching the movie.

Despite the distractibility of teenagers, Hollywood found ways to arrest their attention, particularly in the era of robotics and computer-generated images. *Star Wars, Jurassic Park,* and other movies used special effects to create worlds that existed only on-screen and in the imagination. Movie budgets grew into the millions of dollars, as did the pay of stars. Despite the gulf in earnings between stars and ordinary Americans, brisk tabloid sales made evident the eagerness of Americans to learn about their favorite celebrities, even from sources that were not always credible.

With the rise of the Internet, Hollywood created Web sites for their movies, both current and forthcoming. These Web sites offered movie clips, the equivalent of television commercials, reviews, information about the actors and locale, and other details. Videotapes and DVDs allowed Americans to watch a movie in their own home, making the viewing of a film a private event. Cinema had drifted from the working-class nickelodeon of the early 20th century to the middle-class home at century's end.

RADIO

Radio had its roots in the classical physics of the 19th century. In 1887 German physicist Heinrich Hertz discovered radio waves, a type of low-frequency light outside the visible spectrum. In 1893 Croatian-born American inventor Nikola Tesla demonstrated at a lecture in Saint Louis the transmission and reception of radio waves through space. Italian inventor Guglielmo Marconi in 1895 sent the Morse code by radio waves, raising the possibility of radio transmission of the human voice. University of Pittsburgh electrical engineer Reginald Fessenden did just that in 1906, but the signal was weak and difficult to separate from static. Alert to these problems, Lee De Forest of Armour Institute of Technology in Chicago led the way in the 1910s in improving the vacuum tube to amplify radio signals and reduce static. By 1916 David Sarnoff of the Marconi Wireless Telegraph Company in New York City had decided that the future of radio lay not in sending and receiving private messages. The telephone already served this end. Rather, a radio transmitter would broadcast music, speeches, news, sports, and the like to receivers in homes. In 1918 Columbia University electrical engineer Edwin Armstrong invented the super heterodyne circuit to tune a receiver with precision to the frequency of a radio wave. In 1919 Radio Corporation of America (RCA) formed with the aim of realizing Sarnoff's vision, and in 1920 it gave 10 percent of its stock to American Telegraph and Telephone (AT&T), a move that forged cooperation between the former rivals.

RCA did not at first entice Westinghouse Electric into partnership, and in November 1920 the latter launched KDKA, the first commercial station, in Pittsburgh. In its first hours on air KDKA broadcast the election results that Republican Warren Harding had won the presidency. Subsequent broadcasts were one hour each evening, though with time the program lengthened. Buoyed by the success of KDKA, Westinghouse established stations in New Jersey, Massachusetts, and Illinois. By interesting people in the broadcasts, Westinghouse aimed to create demand for radio receivers, in time known simply as radios.

As it had with AT&T, RCA fashioned an agreement with Westinghouse, giving it in 1921 a million shares of stock and 40 percent of the market for radio equipment. Having absorbed its rivals, RCA became the pioneer of radio. By 1922 RCA was manufacturing radios that produced sound superior in clarity and amplification to the old crystal set and earphones. Mounted in wooden cabinets, the radio was a piece of furniture the family could gather around in the evening for the day's news, though cheap portable radios in the 1930s began to erode sales of the large units.

RCA's broadcast of the 1921 championship fight between heavyweights Jack Dempsey and Georges Carpentier underscored the popularity of radio. From Maine to Florida some 300,000 Americans heard the fight, 100,000 in New York City's Times Square alone. In October RCA garnered even more listeners by broadcasting the World Series between the New York Giants and the New York Yankees.

The appeal of radio led to spectacular growth in the industry. By the end of 1921 radio could count just 10 stations, but by the end of 1922 the number was nearly 350. In 1922 Americans bought $60 million in radios and equipment, and by 1924 the number had leapt to $358 million.

With lots of time to fill, radio stations needed more than sports to keep an audience. News, before the era of cable television, could likewise absorb only a fraction of on-air time, and stations, eager to find a means of filling their hours, turned to music. Classical music was a gamble—since many Americans had never heard a performance, whether live or recorded—but perhaps because of its novelty Americans were willing to listen to classical music, and many hankered for more. Goaded on by what they heard on radio, Americans sought out live performances, and symphony orchestras responded to the demand, growing from 60 in 1928 to 286 in 1939. Schools likewise felt this new demand for classical music. A rarity in 1920, by 1940 30,000 orchestras and 20,000 bands had found a place in America's schools. Radio also turned to jazz, though hesitantly at first for fear that jazz was not quite respectable. Composer George Gershwin's *Rhapsody in Blue* won radio over to jazz, and Guy Lombardo, Duke Ellington, Glenn Miller, and others became household names because of the medium. In addition to classical music and jazz, radio broadcast country music. *The Grand Ole Opry* grew to command four hours a night on Friday and Saturday and had a wide following in the South and Midwest.

In addition to music, radio broadcast religious services and preaching. Stations received letters in praise of religious programming from farmers, the elderly, and invalids who had difficulty attending services. By the 1930s religious programming took on a social and political dimension. The Great Depression led radio personality Catholic priest Charles Coughlin of Royal Oak, Michigan, to attack President Franklin Roosevelt's New Deal for doing too little to uplift the poor.

In 1933 Roosevelt took to the air, in what became known as Fireside Chats, to tell Americans what he and Congress were doing to better their lives. With newspaper editors hostile to Roosevelt, his reelection in 1936, 1940, and 1944 measure, however imperfectly, his success in creating a bond with Americans through radio.

Soap operas, offering a respite from the world of news and politics, captivated radio audiences. None was more popular in the early years of radio than *Amos 'n' Andy*, which had 40 million listeners by the early

1930s. The two protagonists were black men who had moved from Atlanta to Chicago and, along with their sidekick the Kingfish, mangled language in a way that made Americans laugh.

Whatever program they heard, Americans were united in the experience of radio. A housewife in Fargo, North Dakota, and a salesman in Cincinnati, Ohio, might have little in common other than the fact that they heard the same news every evening on the radio. In this way radio knitted Americans into a homogeneous community and diminished the sense of isolation they might otherwise have felt. Radio, in its ability to deliver Americans a world beyond their own immediate one, integrated itself into the lives of those who wanted the vicarious experience of living beyond the familiar. Radio was as much a part of the daily routine as the newspaper and small talk between friends and family.

TELEVISION

Radio had its counterpart in television, which pioneered the broadcast of images. In 1907 British inventor Alan A. Campbell-Swinton and Russian physicist Boris Rosing, working independently, developed television from the cathode ray tube. By 1930 inventors Charles Jenkins and Philo Farnsworth of the United States, Vladimir Zworykin of the Soviet Union, and John Baird of Scotland developed television with finer resolution of images than Campbell-Swinton and Rosing had achieved. In 1928 Jenkins secured the first commercial license for television station W3XK in Wheaton, Maryland, and in 1930 aired the first program. In 1939 RCA broadcast President Roosevelt's speech at the World's Fair, an event that received international coverage. By 1948 the United States had forty-eight television stations in twenty-three cities.

After World War II Americans, awash in prosperity, bought televisions, sometimes more than one for a home. Television competed with cinema, which in 1946 peaked with some 90 million viewers a week. Within two years, television was eroding cinema's audience. In 1948 Americans had 250,000 television sets, and in 1952 17 million sets. During these years the three networks arose: American Broadcasting Company, Columbia Broadcasting System, and National Broadcasting Company, all of which had begun as radio broadcasters. The networks expanded several of their radio programs into television. *Gunsmoke, The Guiding Light,* and *The Twilight Zone* were so successful as television shows that many Americans forgot they had begun as radio programs. Whereas filmmakers fought television in the 1940s, refusing in 1948 to release movies to television, by the 1950s cooperation was more common. In 1954 Disney wrote *Disneyland* for television and in 1955, *The Mickey Mouse Club.*

By the mid-1970s half of American homes had at least two televisions. In buying the television in such quantities, Americans could scarcely avoid its shaping their lives. The television became the technology around which families arranged the furniture in the living room and in the bedroom in instances in which they had more than one television. The focal point of communication and entertainment, as radio had been before the war, television shaped the routine of daily life. Those most smitten by the technology waited until a commercial to use the bathroom and felt anxiety upon missing an episode of their favorite program. Families that had gathered around the table for dinner now curled up on a couch to watch television while they ate a TV dinner or a bowl of chips. Snacking while watching television became commonplace and may have contributed to the obesity of many Americans. In homes with more than one television, family members could watch different programs on televisions in different parts of the home, contributing to a sense of isolation in families that were not careful to preserve activities that brought their members together.

In her book *The Plug-in Drug,* Marie Winn emphasized the addictive power of television. One could watch programs hour after hour in something akin to a stupor, expending little energy and less thought. The television remote control allowed Americans to flip mindlessly through channels in a random quest for a program that would hold their attention. The sequence of images in a program and a succession of one program after another, if not random, could seem so when viewed en masse. By 1963 Americans watched television almost six hours a day, nearly as much time as they spent at work—a measure of television's narcotic effect.

From the outset television shaped how Americans perceived the family and gender roles. The popular *I Love Lucy* stereotyped Lucille Ball as a housewife, one with unattainable dreams and a modest intellect. Her husband in life and on television, Desi Arnaz, earned the couple's livelihood in the show. Their marriage stable, the characters Ball and Arnaz played never contemplated divorce or an extramarital affair. As conventional as *I Love Lucy* was in these respects, it gave white America a mixed marriage, for Arnaz was Cuban. He was also unconventional in being a bandleader rather than a nondescript corporate employee. His heritage led one television producer to fear that Arnaz would offend the sensibilities of white viewers. To the contrary, 10.6 million households watched *I Love Lucy* in 1952 and nearly 50 million in 1954. These audiences watched Ball challenge the conventions of television. The television industry routinely denied roles to pregnant women and banished the word *pregnancy* from dialogue. Ball, however, worked through her pregnancy and, when the baby came in January 1953, 44 million people watched the show, twice the number as watched President Dwight D. Eisenhower's inauguration the next day.

Mirroring the complexity of the United States, television in the 1950s brought viewers the buoyant energy of Elvis Presley and the malevolent bullying of Wisconsin senator Joseph McCarthy. Television welcomed Presley but took on McCarthy at the height of his power. Journalist Edward R. Murrow televised *A Report on Senator Joseph McCarthy* in March 1954, exposing McCarthy's lack of scruples in attacking his enemies. That same year television brought Americans the Army-McCarthy hearings, further discrediting McCarthy and goading the Senate to censure him.

The McCarthy hearings demonstrated television's power to bring politics into the lives of Americans. Confirmation of this power came in the televising of the 1960 presidential debates between John F. Kennedy and Richard Nixon, which some 65 million Americans watched. Americans who heard the debates on radio gave Nixon the edge, whereas television viewers thought Kennedy had won. Kennedy's narrow victory in November 1960 seemed to confirm that Americans turned to television rather than to radio in forming an opinion of their leaders. In 1963 Americans watched Kennedy's somber funeral. Television also brought into America's living rooms the civil rights movement, the Vietnam War, and the antiwar movement.

Like politicians, religious leaders used television to affect the lives of Americans. Catholic priest Fulton J. Sheen was popular enough to win an Emmy Award in 1952 as television's outstanding personality. With a television audience of 25 million he at one point approached *I Love Lucy* in ratings. Billy Graham used television in the last half of the 20th century to build a ministry with popular appeal. Graham and the evangelists who followed him attracted Americans with an uncomplicated message of redemption. Americans simply needed to accept Jesus into their lives to gain salvation. Television evangelists for the most part ignored social justice and made no ethical demands of their viewers. Using television to advantage, evangelists satisfied Americans by giving them simple answers to difficult problems and by making complacency a virtue. Americans responded to this approach. Whereas fewer than half of Americans belonged to a church in 1940, the percentage had risen to 73 by the late 1950s.

Some evangelists railed against violence and sex on television, and reformers in the 20th century periodically attempted to sanitize what Americans, particularly children, watched. In the 1970s the National Association of Broadcasters touted Family Time, the televising of inoffensive content between 7 and 9 P.M., when families would presumably watch television. Family Time rested on the belief that most children went to bed by 9 P.M., at which time the airwaves were opened to adult content. The Writers Guild of America challenged Family Time as censorship, underscoring the tension between two visions of television: television as a free market of content and television as purveyor of values.

Cable television arose in the 1970s to challenge the networks, giving Americans a choice among a bewildering assortment of programs. The rise of the cable station MTV targeted adolescents with a barrage of images—with the music videos' images of scantily clad women geared especially toward adolescent boys. MTV popularized music videos in which viewers watched a rapid sequence of images with abrupt transitions between images. CNN televised news around the clock, and by the end of the 20th century cable news networks arose that did not so much broadcast news as bundle it with an ideology. Cable television also pioneered the reality show. Rather than use actors, the producers of reality television filmed people in unscripted dialogue and action. Viewers did not seem to mind that the producers skewed reality by filming attractive people rather than the homely or in coaching dialogue that was to be unscripted.

Two technologies broadened the appeal of television. The first, the VCR, allowed the technophile to program a VCR to tape programs for viewing later. No longer did the network schedule dictate when Americans watched a program or ate a meal. Perhaps the most popular use of the VCR was in allowing Americans to watch movies that had been at the theater months earlier. The second was the video game. From a simple beginning as video tennis, video games in the late 20th century grew to have a lifelike quality. Full of violence and the titillation of sex, video games sparked controversy but nonetheless commanded an audience of gamers—players who lived in the artificial construct of video games rather than in the reality of the workaday world. At the end of the 20th century, television had only the automobile, the computer, and the cell phone as rivals in its power to transform American life.

THE COMPUTER AND THE INTERNET

The development of the personal computer opened a new era in communication in 20th-century America. In 1971 Intel engineer Marcian "Ted" Hoff invented the microchip, making possible a reduction in the size of computers. The new computers could fit atop a desk and were suitable for business and home. In 1976 American entrepreneurs Steven Jobs and Stephen Wozniak founded the company Apple and introduced Apple 1—a standardized, inexpensive personal computer—on the principle of Fordism: create a product for a mass market to tap into American consumerism. In 1983 Apple unveiled Lisa, the first computer with icons and a mouse to navigate among icons. The Macintosh supplanted Lisa in 1984 at a cost of $2,500. In 1987 the Mac II became the first computer to have a color printer. International Business Machines (IBM), Apple's rival, introduced in 1981 its first personal computer.

But the success of computers depended on software. In 1975 American entrepreneurs William Gates and Paul Allen founded Microsoft in Redmond, Washington, and licensed its software to IBM and in the 1980s to Apple. Ultimately, though, the power of computers lay in joining them in a network. As early as 1962, Massachusetts Institute of Technology psychologist Joseph C. R. Licklider had envisioned a universal network of computers. In 1969 the Defense Department's Advanced Research Projects Agency created the first network among the agency's employees and academicians. The network grew in the 1980s to include the National

Science Foundation, NASA, the National Institutes of Health, and government agencies in Europe. In 1990 engineer Tim Berners-Lee at Organisation Européenne pour la Recherche Nucléaire in Switzerland devised the World Wide Web, which used hypertext markup language, a simple coding language, and universal resource locators, an address for each Web page. A person navigated the Web by a set of rules Berners-Lee called hypertext transfer protocol.

By 1999 half of all American homes had a personal computer and one-third had an Internet connection, whereas in 1994 the numbers had been only 27 percent and 6 percent, respectively. The use of computers and the Internet on a large scale shaped the lives of Americans in an incredible variety of ways, though e-mail was perhaps the primary reason Americans used a computer. According to one 1999 survey, 67 percent of adolescents aged nine to seventeen used a computer for e-mail, compared with 58 percent for games and 50 percent for homework. Adolescents found e-mail indispensable in expanding their network of friends, and both adolescents and adults used e-mail to span distance much as the telephone did in the 19th and 20th centuries. Parents e-mailed their children who were away at college, and relatives one another across time zones. Spouses separated during travel found a simplicity and intimacy in their e-mails that were hard to duplicate by phone. Photo attachments announced the birth of a child or the exuberant moments of a vacation in Florida. Instant messaging enhanced further the possibilities of communication by computer.

Chat rooms were yet another forum in which people came together in virtual space, forming a sense of community. Americans used this forum to share interests, finding that they had more in common with a person in another state than with their neighbors. Some cultural observers wondered whether Internet communities were replacing the family as family members went their separate ways on the Internet. Particularly worrisome were children who entered chat rooms on the pretense of being adults and adults who entered chatrooms on the pretense of being children.

Americans who wanted more than friendship posted a photograph and personal information on dating sites. Match.com, founded in 1995, had by 1999 1,400 members who claimed to have married a person they met on the site. These numbers suggest that at least some Americans found the experience of meeting someone online less fraught with anxiety than meeting a person at a bar or on a blind date. Nevertheless, an online encounter was not risk free. The anonymity of the Internet made it easy for married men and women to feign being single, and people who began an online romance did not always live up to expectations when they met in person.

Americans who were concerned about their health consulted Web sites about a medical condition or surgical procedure. By one estimate the Internet had in 1999 20,000 Web sites devoted to some aspect of health. Beyond researching a medical condition, Americans used the Internet to fill prescriptions, to order medicine for their pets, and in some instances to e-mail their doctor. E-mail correspondence between doctor and patient blossomed as doctors grew accustomed to using the Internet. Whereas only 5 percent of doctors had used the Internet as part of their practice by 1996, 85 percent had by 1998.

The lure of money drew other Americans online. Investors used the Internet to track and trade stocks. By 1998 Americans traded one of every seven stocks online, with the most aggressive traders buying and selling more than 100 stocks a day online. This volume of trading sometimes led to errors in which an investor keyed in "1,000" when he had intended to buy only 100 shares. Others used the Internet to calculate how much money they would need to save for retirement or for their child's college tuition. Still others did their banking online, transferring funds or setting up their bills to debit a checking account.

Job seekers used the Internet to expand their search beyond newspaper classifieds, though in apparent redundancy some newspapers posted their classifieds online. Companies, universities, and all levels of government likewise posted job openings online. In 1999 Monster.com, a Web site for job seekers, posted 210,000 jobs and 1.3 million résumés, with job hunters adding another 40,000 résumés each week. That year, 82 percent of students graduating from college looked online for work. Beyond searching for work, job seekers researched companies, cost of living, and salaries, sent resumes by e-mail to employers, and applied for work through company Web sites.

The Internet changed how Americans pursued their education, in part by freeing the student from the need to be in the same room as the instructor. This update of the old forum of correspondence classes, distance education used online syllabi, course outlines, exam questions, and e-mail in lieu of the traditional lecture. Stay-at-home parents took courses toward bachelor's degrees. Recent college graduates, looking to upgrade from their first job, worked online toward advanced degrees or supplemental classes and, in an informal setting, retirees joined online study groups. Colleges accepted applications online and prospective graduate students took the Graduate Record Exam on a computer.

The Internet changed the experience of shopping for Americans who chose not to fight traffic and wait in line in a store. Shoppers could buy everything from books to cars in a virtual marketplace full of inventory. Eager to attract customers, businesses—some without storefronts—created Web sites from which to sell their products. Sears and other retailers offered the same merchandise online and in its stores. With large chains such as Sears, geography was not a barrier to in-store purchase, but singles stores, such as American Girl in Chicago, would, for example, want to be sure that a girl in Salt Lake City who wanted the doll Molly was able to purchase it either by phone or at Americangirl.com. Amazon.com was the equivalent of a virtual mall in the breadth of its offerings, and eBay

In the late 20th century computers were ubiquitous among the middle class. The laptop made the computer portable, allowing Americans to use it virtually anywhere. Courtesy: Shutterstock.

provided a marketplace for Americans who wanted to auction their belongings. In 1998 Americans spent $11 million online, and even more when one accounts for store purchases of an item first seen online.

The Internet also transformed the experience of travel, making the travel agent redundant. Americans using Travelocity could book both flight and hotel. By 1998 more than 6 million Americans made their travel plans with Travelocity, spending more than $285 million on airline tickets and hotel rooms. By 1999 Travelocity had access to 95 percent of the world's airline seats, more than 40,000 hotels, and 50 car rental companies. Americans could also use Travelocity to book a cruise to the Mediterranean, the 1, or some other exotic locale.

E-mail

One wonders how many people have invented new personas and life histories when going online. Some married adults declared themselves single, and some men pretended to be young boys in hopes of making predatory and illegal overtures to a young boy or girl. Still others liked the anonymity that the Internet afforded. With no consequences for their behavior, sexists and racists used the Internet to spew hatred.

THE TELEPHONE

Alexander Graham Bell invented the telephone in 1876 and formed Bell Telephone Company the next year. By 1900 New York bankers rather than Bell controlled the company, now named American Telephone and Telegraph Company. AT&T operated as a series of state-based telephone companies, and state regulation shielded AT&T from competition. Under the leadership of Theodore Vail, the company bought its smaller rivals, and for a time AT&T came as close to a monopoly as one might imagine. Vail envisioned an America in which everyone had a telephone, just as Henry Ford envisioned universal access to the automobile. Vail's vision remains elusive. The first to buy a telephone were high-income professionals, much as the wealthy were the first to buy an automobile. In 1946 half of all homes had a telephone, but most of these were in the cities. In the countryside less than one-quarter of homes had a telephone. To spur the growth of telephones in the countryside, Congress in 1949 authorized the Rural Electrification Administration, a holdover from the New Deal, to offer telephone companies low-interest loans to expand coverage beyond the city and suburb. Blacks lagged behind whites in telephone ownership: in 1989 more than 90 percent of white homes had a telephone, but only 85 percent of black homes had one. By 1996 94 percent of American homes had a telephone, and 40 percent had an answering machine.

The telephone shaped daily life by the sheer magnitude of its use. In 1986, for example, Americans made more than 350 billion calls, 90 percent of them local, and averaged twenty to thirty minutes a day on the phone. Originally a tool of business, the telephone linked homemakers who otherwise would have been isolated from one another. Homemakers and young singles used the telephone the most, followed by teens. Men and seniors over age 75 used the telephone the least. One-quarter of all calls were for the purpose of chatting or socializing. One-fifth solicited or dispensed news and information, and the rest coordinated activities, resolved problems, or conducted business.

Among other Americans, teens used the telephone as a social technology discussing by phone all the issues pertinent to 20th-century teens. But the teen who spent hours most nights on the phone risked a confrontation with siblings and parents, all of whom wanted their turn. The obvious solution of installing a second and perhaps even a third telephone line conflicted with the need of many families to economize.

Teens used the telephone to forge friendships, to gossip unabashedly about other people, and to set up dates. Throughout much of the 20th century, the ritual of dating assigned the boy the privilege of calling a girl for a date. For her part, she was to be submissive in awaiting his call without complaint. Once established as a couple, the boy called his girlfriend to chat, an informal means by which the two solidified their

In the 20th century men used the telephone less than women and teens. Nevertheless some men, adamant that they paid the phone bill, regarded the telephone as their property. Courtesy: Library of Congress.

relationship. The sexual revolution of the 1960s empowered women in many ways. One was to embolden girls to take the initiative in setting up a date by calling a boy. No longer was she passively to await his call.

In using the telephone as a social technology, homemakers, like teens, gossiped about others. In addition, homemakers exchanged recipes with other homemakers, offered encouragement to friends who were having problems, and expressed their aspirations for the future. Advertisements aimed at women, aware of how important the telephone had become in the lives of homemakers, declared that no home was complete without a telephone.

Coupled with the telephone was the answering machine. Americans bought an answering machine to catch an important call while they were out. Not satisfied with this utilitarian function, Americans also used the answering machine to screen calls. Telemarketers and bill collectors rang through to the answering machine; but when a friend called, the person who owned the answering machine picked up the phone so the two could chat. In such a circumstance, the friend knew that the other person was screening calls. Caller ID, however, made it possible to screen calls without the caller knowing. The person who had caller ID simply checked the number of the incoming call and let undesirables ring through to the answering machine. Call waiting allowed one to take an incoming call while putting the first party on hold, an act that my have seemed rude to the person who had been a party to the first phone call. Call waiting gave the recipient of the call a convenient excuse for ending the first call if it

The telephone eased the isolation of housewives. They used the telephone to gossip, to exchange recipes and to offer emotional support to one another. Courtesy: Library of Congress.

had run its course. If the second call, on the other hand, was from an undesirable caller, the recipient let the call ring through to the answering machine.

The telephone shaped even the lives of the homeless. Being poor, they had no telephone, and without one they were difficult to track down by potential employers. In this sense, the lack of money, a phone, and a job all conspired against the homeless person. An agency in Seattle, Washington, sought to improve this circumstance by giving its homeless clients access to free voice mail. Voice mail cut the time in which a homeless person might find a job from several months to just a few weeks.

THE CELL PHONE

Like the telephone, the cell phone permits two people to communicate over distance. Here the similarity with the telephone ends, because the cell phone is more like radio. The cell phone uses radio waves at frequencies not used by radio transmitters to carry the human voice. Because radio waves are everywhere in space, the cell phone need not be tethered to a room. It can roam with the person using it, giving the cell phone its portability.

In 1947 Bell Laboratories first harnessed radio waves for a two-way conversation, in this case creating the police walkie-talkie. That year AT&T, foreseeing the commercial potential of a portable phone, petitioned the Federal Communications Commission (FCC) to set aside a bandwidth of radio waves for these phones. The FCC, however, saw no urgency since no one had yet developed a cell phone, and only in 1968 did it grant telecommunications firms a segment of radio waves. Even then, Martin Cooper of Motorola would need another five years to develop the first cell phone, but at two pounds and $4,000 it was neither portable nor affordable. Motorola and its competitors worked to miniaturize, lighten, and cheapen the cell phone, as was the trend with the personal computer, and in the 1980s subscribers began to gravitate toward the device. Between 1985 and 1987, the number of cell phone subscribers leapt from fewer than 350,000 to more than 1 million, and by 1999 to more than 100 million.

Before the cell phone, a person stayed put in a room to use the telephone. In this respect the telephone was like television and the personal computer in confining the user in space. The cell phone cut the chord of dependence, freeing the user to carry this technology wherever he or she went. Yet portability had pitfalls. Whereas the telephone user simply went outdoors or in his car to be free of the unwelcome caller, the cell phone made its user available to all callers anytime, making it harder to evade a call. The user who did not wish to be bothered had to turn off the ringer or not answer the call. Even when the call was a welcome one, it might come at an inopportune moment, forcing the user to decide whether to talk at length or attend instead to the matter at hand.

In allowing its user to talk by phone just about anywhere, the cell phone empowered its user to keep in closer contact with friends and family than the telephone permitted. Adolescents might have their own cell phone to talk with friends, but this privilege sometimes came at the price of having to call their parents periodically to give updates on their activities and renew their promise to be home by 10 P.M. A spouse could let her mate know that she was running an errand after work and so would not be home before 6 o'clock. A person with a flat tire on his car could with his cell phone call the local branch of the American Automobile Association for assistance, or a coach could call a parent from the field to ask him or her to bring a few extra blankets to the game if perhaps the temperature had dropped faster and farther than anyone expected. Beyond these benefits, Americans used the cell phone in their intimate lives, increasing the odds of two people connecting.

Despite all it could do, the cell phone had limits. Its use at funerals and high-end restaurants branded its owner insensitive and impolite. Amtrak set aside a quiet car on which passengers could be free from this nuisance. New York City, responding to the yearning for silence, banned the use of cell phones in theaters and museums. Hospitals and doctor's offices routinely banned the use of the cell phone on the grounds that it interfered with their equipment.

Even in silence, the cell phone enlarged the experience of daily life. Like the computer, many cell phones could send and receive text and photographs, and even connect to the Internet, making the Internet mobile, an achievement the computer could not duplicate. As the automobile had done with people at the beginning of the 20th century, the cell phone made communication mobile at century's end.

Users and Abusers

The technologies of communication were a mixed blessing. They gave Americans unprecedented speed of communication, but this speed came with a price. The telephone helped siblings living in other states keep in touch, but it was also the technology of nuisance calls. E-mail became a primary means of communicating, but one had to be careful whom one e-mailed. A work colleague might forward in hostility what was to have been a private message to a group of colleagues, or a single e-mail might trigger a large number of responses, not all of them from friends. Radio and television had plenty of choice in programming, but interest groups took offense to some of the programming, objecting in particular to television programs with action suggesting or laden with sex, violence, or profanity. Despite these problems and malcontents, the technologies of communication had a secure place in the lives of 20th century Americans.

7

SCIENCE, TECHNOLOGY, AND THE MILITARY

The Constitution empowered Congress to raise an army and a navy. The U.S. Armed Forces were eager to supplant manpower with technology, and the history of military technology in the United States has been one of increasing lethality. Soldiers, who in an earlier era might have killed face-to-face, in the 20th century killed at a distance, with rifles, machine guns, land and naval artillery, and aerial bombardment. The prospect of war required even more lethal technologies as nations competed in an arms race. Armies did not need to use technology for it to be a deterrent: after the first use of the atomic bomb in World War II, nations, including the United States, have since refrained from using nuclear weapons.

In the nuclear era prominent scientists have debated the consequences of a nuclear exchange between nations. Thousands, perhaps millions, would die. Those not killed by the initial blast would succumb to radiation poisoning. The blast might open a hole in the ozone layer, leaving all life vulnerable to large doses of ultraviolet radiation. Terrestrial life might succumb to ultraviolet exposure, leaving only life in the ocean. Military technology has evolved to the point that it cannot be used without catastrophe.

TECHNOLOGY AND MILITARY LIFE IN THE EARLY 20TH CENTURY

Technology had by 1900 given soldiers arms more lethal than any in the past, but it had done little to ease the rigors of daily life. The Revolutionary

War soldier would have recognized much in the drudgery of soldering in the first years of the 20th century. Before the invention of the truck and the jeep, only the train spanned distance. Otherwise a soldier marched from place to place. Once he had gathered his compass, maps, ammunition, canned rations, and other items into his pack, he found himself weighed down by 70 pounds. Marching fully laden for several hours at a time left American soldiers, no matter how fit, in a perpetual state of fatigue. At the end of a long march, they dug a foxhole and cared for the horses and mules in their company. Before the invention of the artillery trailer, horses were needed to drag artillery, and before the invention of the truck, mules carried the supplies that soldiers did not lug in their packs. Douglas Mac-Arthur recalled as a young officer the care of horses and mules as an onerous task. Many other soldiers agreed.

Once they were spared long marches and the care of equines, American sailors nonetheless did not escape unpleasantness in their daily lives. Steam engines powered ships in the early 20th century, and these consumed vast quantities of coal. From the dock, sailors shoveled coal into large canvas bags. Hoisting the bags aboard ship, sailors dumped the coal into wheelbarrows, which they muscled to manholes and dumped down chutes to the boiler room. This work enveloped them in a cloud of coal dust that covered skin, clothes, and food. The dust settled into wrinkles in the skin and into the crevices around the eyes, where it was difficult to remove. The boiler room, where the stokers shoveled coal into the furnace, was so hot that they worked naked except for wooden clogs on their feet. The loss of salt and fluids from this work caused dizziness, nausea, and vomiting, and the task of cleaning the boiler room stirred up a fog of asbestos, dust, and soot, prompting sailors to wear protective masks. Before the discovery of DDT, astonishingly large cockroaches were plentiful aboard ship. For amusement sailors raced them across tables in the mess hall, placing bets on which one was fastest. Rats were likewise part of the shipboard fauna. Before the discovery of the antibiotic streptomycin and the development of the tuberculosis vaccine, the tuberculosis bacillus spread easily among sailors living in close quarters, circulating in the poorly ventilated air of the ship. For light, sailors lit oil lamps, which put grime, and soot into the air and required daily cleaning.

Life aboard a destroyer was especially difficult because of its small size. In rough water a destroyer rolled as much as 50 degrees from perpendicular. Being buffeted to such an extent, sailors could not eat from their tables and only with difficulty slept in their bunks. In this environment, their normal routine was enough to bring on exhaustion. Life aboard a ship, whatever its size, was crowded, noisy, hot in summer, and cold in winter. Twenty-eight sailors shared a single sleeping compartment fifteen feet wide, eight feet tall, and twenty-five feet long. Each bed, separated from its neighbor by twenty-five inches, was a steel frame with a two-inch mattress. A sailor who weighed around 200 pounds would sag into the breathing space of the person below him, forcing that sailor to sleep on his side or stomach. Officers allowed sailors fresh water for bathing only at intervals, forcing them to go days, sometimes weeks, without a bath. Finally, sailors spent twelve hours a day at their guns, whose repeated firing caused partial deafness.

CHEMISTRY, WAR, AND LIFE IN THE ARMY

Among the problems chemists set themselves during the 19th century was the derivation of poisons to kill everything from houseflies to humans. Research yielded the gases phosgene and chlorine and the vaporized liquid mustard gas. By century's end, the armies of the United States and Europe had added these chemicals to their arsenals. In 1899 conferees at the Hague Conference voted to ban chemical warfare, but the United States dissented on the grounds that an army was no more barbaric to kill an enemy with mustard gas than with a machine gun. The United States claimed that to ban chemical poisons but permit conventional armaments was capricious.

The nicety of treaties aside, the real objection to the use of chemical poisons was tactical. The wars of the 20th century, theorists believed, would hinge on speed. Artillery would support cavalry and infantry in a lightning strike deep into enemy territory. Wars of movement, modern combat would not fix large numbers of troops in place, and so they would not be vulnerable to a chemical attack. A war of mobility in its first weeks, World War I defied military planners by devolving into a war of position. Acknowledging that this static war suited the use of chemical poisons, French troops in August 1914 lobbed poison grenades at the Germans, though

World War I was fought with chemical poisons in addition to traditional armaments. Although it was uncomfortable to wear, the gas mask protected soldiers against a gas attack. Courtesy: Library of Congress.

the gas dissipated without killing anyone. A German attack that October was equally ineffective, but in April 1915 the Germans used mustard gas with ghastly results at Ypres, Belgium, ushering in the era of chemical warfare.

Upon America's entry into the war in April 1917, every soldier was trained in chemical warfare. Instructors made the grim observation that during a gas attack, soldiers fell into two categories, "the quick and the dead." Soldiers donned masks in simulation of a gas attack. These they wore in their foxholes, in setting up a machine gun, in dashing across the land as though breaching the enemy's position, and while marching. These exertions had predictable effects: the lens on the goggles fogged with perspiration, and one lieutenant could not stop drooling because of the mouthpiece, his saliva dripping down the flutter valves and onto his shirt. Private Norman Dunham complained that his mask was "the worst thing a soldier had to contend with" and "the most torturous thing" a soldier could wear. Training culminated in immersion in a gas chamber, a soldier wearing a mask and exposed to tear gas for five or ten minutes. Some troops endured even more thorough training over three days, the most intensive the Army offered. Medics wearing masks carried stretchers with simulated wounded men. Engineers wearing masks built roads, and some infantry dug ditches while wearing masks. The ordeal lasted hours; afterward, these troops played baseball while still wearing their masks.

Once chemical warfare training was complete, troops visited hospitals near the front to see the casualties of poison gas to impress upon American soldiers the reality of chemical war and to dispel the rumors about a chemical attack's causing the body to lose its eyes, fingers and toes, though those rumors persisted. When the first American troops entered the trenches in January 1918 to relieve a French company, they witnessed the French evacuating 200 poison gas casualties, a more sobering episode than the hospital tour. The French posted instructions for what they called green troops, requiring soldiers, both French and American, to wear gas masks for four hours after a gas attack. Believing this rule arbitrary, American commanders allowed junior officers to tailor a decision to the circumstances on the field. An officer took a risk in pulling off his mask, but he did so just long enough to detect the odor of gas. As long as he smelled gas, his troops kept on their masks. In some cases day and night attacks prompted soldiers to keep their masks on for eighteen hours a day. Officers made other efforts to protect themselves, covering their dugouts with blankets to shield against poison gas. Inopportune winds could, however, open a dugout to gas, imperiling the officers inside. To alert troops of a gas attack, a sentry blew a horn or rang a bell, but this alarm was not always enough. In February 1918 a German gas attack killed eight Americans and wounded seventy-seven others. Thereafter Americans and Germans shelled one another with poison gas at every opportunity. American soldiers feared mustard gas most of all for its tendency to burn skin. The vaporized liquid could blind victims and irritate the lungs, in the worst cases causing the lungs to bleed, a usually fatal condition. Seeking protection, Americans covered their faces and hands with a thick layer of Vaseline to prevent mustard gas from reaching their skin, but mustard gas embedded itself in the Vaseline, and thus the removal of the petroleum gel required care. Officers issued their charges fans to clear an area of poison gas, but this was a futile gesture. When soldiers had done all they could to protect themselves, they masked their horses and mules.

Soldiers who were unlucky enough to be victims of a gas attack—as at Belleau Wood, where mustard gas blinded some soldiers and left others crying and delirious with pain—were pulled back to a field hospital. Those who could stand showered to remove any gas residue. Medics bathed those who could not stand, taking care to spray the eyes, nose, and throat with bicarbonate of soda to neutralize any reaction between the gas and sensitive membranes of the body. Soldiers breathed pure oxygen and were sometimes bled, a holdover from ancient ways of treating the ill. Those who had taken food or water contaminated by gas ingested cod liver oil or castor oil to coat the stomach lining. Soldiers who showed no improvement were evacuated to a base hospital that specialized in the treatment of gas victims.

THE MACHINE GUN AND LIFE IN THE TRENCHES

In the 19th century four different Americans claimed to have invented or improved the machine gun. Units of the army used it during the Civil War and the Spanish-American War, but the army officially adopted the machine gun only in 1904 when, following tests of several models, the chief of ordnance chose the Maxim water-cooled machine gun, the latest model by the inventor Hiram Maxim. The Maxim gun had the advantage of not overheating, though it was heavier than an air-cooled gun. But reports that the Maxim gun jammed frequently led the army in 1909 to replace it with an air-cooled French model, the Benet-Mercier. At twenty-two pounds it was lighter than the Maxim gun, but it tended to overheat. Firing the Benet-Mercier in short bursts avoided the problem of overheating, however, and it remained in service even as the army sought in 1914 to replace it with a German model, the Vickers water-cooled machine gun.

From the outset, the army defined the machine gun as a defensive weapon. Even the relatively light Benet-Mercier was too heavy to carry in a sprint into enemy territory. Nor could one fire it accurately from a standing position: better to set a machine gun on a tripod and the tripod on the ground. Pivoting atop a tripod, the machine gun sprayed bullets across a broad field of vision. Set at intervals along a line of soldiers, the machine gun protected against a frontal assault.

By making the frontal assault a type of suicide, the machine gun reduced the initial mobility of World War I into stasis. In France and Belgium the French, British, and Belgians on one side and the Germans on the other dug in at the front, transforming their positions into long lines of trenches, which the Americans used upon entering the war.

American soldiers found life in the trenches to be unpleasant. They were exposed to rain and snow and had difficulty keeping warm in winter. Absent an enemy attack, they had little to do and nowhere to go. At night small detachments of men probed No Man's Land for breaks in the barbed wire while the rest slept in holes they had dug in the trenches. Lacking entrenching tools, some soldiers dug these holes with bayonets. In winter they swung pickaxes to gouge holes out of the frozen soil. Lice (called cooties) and rats were ubiquitous in the trenches, as was the smell of excrement and disinfectant. Artillery attacks exacerbated the misery; a direct hit, soldiers knew, was fatal. The Marines at Belleau Wood made explicit the dangers of life in the trenches by calling their trenches graves. Fresh water was often in short supply and a hot meal nonexistent. The Marines at Belleau Wood had nothing to eat but canned corned beef or raw bacon and biscuits. Troops more fortunate than these Marines foraged in gardens for potatoes and carrots.

As American soldiers moved into the trenches, their horses brought up the artillery. Whenever the artillery moved, soldiers had to dig a new pit for each piece, a bunker for the shells and powder charges, and foxholes for themselves, all in hopes of lessening their vulnerability to attack by enemy artillery or an airplane. Repeating this drudgery ad nauseam, artillerymen endured lives of unremitting toil.

During the war, the growing number of trucks and artillery trailers allowed soldiers in some units to escape the chore of caring for horses and mules. The trucks that replaced mules were not at first comfortable. With solid rubber tires and no suspension, these trucks jolted drivers and riders alike. Moreover, a ride in a truck, though a respite from marching, meant that troops had to get somewhere in a hurry, often to the front.

In spring 1918 the first soldiers came down with influenza. The outbreak was modest, and no one gave much thought to the possibility of a pandemic when, in the fall, soldiers again became ill. Those who had had the flu in spring were partially immune to its iteration that fall. Army physicians vaccinated soldiers as a precaution, but the vaccine did not confer immunity to this strain of influenza. In any case, the army would not likely have had enough inoculum for everyone in uniform. Without a vaccine, the Army fell back on quarantining the sick, inspecting soldiers for the first symptoms of flu, ordering windows closed in barracks, putting up sheets between beds, airing out bedding daily, and, in one instance, draping a sheet down the center of tables in the mess hall. Those with flu received bed rest, aspirin, blankets to keep warm, and a light, hot diet.

These measures did not banish death, however. At war's end, the army reported 46,992 deaths from flu, and the Navy 5,027.

THE AIRPLANE AND LIFE IN THE MILITARY

The nation that had invented the airplane, the United States, found itself lagging behind the combatants in World War I. In April 1917 the United States had just twenty-six pilots and fifty-five aircraft, fifty-one of which General John Pershing, commander of the American Expeditionary Force, declared obsolete. By comparison, France had 1,700 planes. The plight of the U.S. Air Corps was even worse when one considered that the armies involved in the war had millions of men. What could fifty-five planes do against enormous, well-entrenched armies?

Nonetheless, the U.S. public in the United States saw the aviator as a lone warrior battling for country and with courage against a foe who determined to shoot him from the sky. From the beginning, the air war had a romance about it that was absent from the nameless mass of men who fought in the trenches. The triumph of America's flying aces, Eddie Rickenbacker among them, made headlines.

But the reality of air combat was not as romantic as the person on the street believed. The open cockpit offered the pilot no protection against the elements. The higher he flew, the colder the air, and frostbite quickly set in. In hopes of keeping the wind at bay, he covered his face with a thick coat of Vaseline, but without an oxygen mask until 1918, an aviator risked dizziness and even blackout at high altitude. Engine oil spattered the face of aviators, obscuring their vision. The dangers of flying were evident in the statistics: more American airmen died in crash landings than were shot out of the sky.

On the ground, American troops viewed the airplane as an intrusion in the routine of combat. Artillery and the machine gun were an organic part of the technology of war, but the airplane, strafing Americans from above, did not play fair. American pilots made matters worse on occasion by strafing their own troops by mistake. American troops responded in kind, sending up a barrage of bullets at the sight of a plane, no matter its nationality. Matters became so serious that Pershing's staff issued a leaflet to every doughboy forbidding him from shooting at an American plane.

Between the world wars, engineers converted the biplane to the monoplane, encased the cockpit in glass, increased the speed and, thanks to the oxygen mask, increased the altitude to which an airplane climbed. Planes could now do more than strafe troops; they could carry bombs and torpedoes. Launched against a city, aircraft could bomb a factory or oil refinery. Launched from the deck of an aircraft carrier, they could bomb or torpedo a battleship, as the Japanese did to the American fleet at Pearl Harbor. Despite these advances, life aboard an airplane in World War II was far from luxurious. The B-17, America's bomber in Europe, was unheated and unpressurized, requiring heavy clothing and the ubiquitous oxygen mask. A gunner whose machine gun jammed risked frostbite by removing his gloves to fix it. On occasion, the electric wire that heated his suit snapped, again putting him at risk of frostbite.

After World War II, engineers replaced the propeller with the jet engine. Life in the cockpit of a jet had the feel of running a big rig for miles along a highway. Flight times changed little, and even the targets in Korea and Vietnam were reshuffled so that a jet pilot would be able to fly the same route several times. Antiaircraft fire took a toll against the jets, particularly when they flew at low altitude in support of ground troops. In response the U.S. Air Force proposed two types of aircraft: a jet that flew high enough to be out of range of antiaircraft guns and a jet that evaded radar. The first was the B-52 bomber, which had a ceiling of 50,000 feet. The Stratofortress had eight jet engines, a speed of 650 miles per hour, and a payload of 70,000 pounds. The second, the F-117A Nighthawk, with its angular shape, resembled a bird when it appeared on a radar screen. The facets on the Nighthawk deflected radar waves and its coating absorbed some radar waves, further reducing the number of waves that returned to the detector. The Nighthawk also minimized its emission of heat, an energy that an enemy could track, by mixing cold air with the exhaust.

THE AIRCRAFT CARRIER AND LIFE IN THE NAVY

American naval theorist Alfred Thayer Mahan believed a nation's power to influence events resided in its navy. Mahan called on the United States to build a big fleet with which to challenge potential rivals. To admirals at the end of the 19th and beginning of the 20th century, the battleship—with its guns, armor deck, and layers of steel plates—was the great projector of naval power. A navy was strong as long as its battleships had uncontested control of the seas. All other ships were to serve the needs of the battleship: cruisers and destroyers for protection against surface vessels and submarines to guard against an undersea threat.

The airplane threatened to undermine the supremacy of the battleship. Eager to adopt the airplane, the U.S. Navy leapt from 7 aircraft in 1914 to 2,100 in 1918. In 1925 army general Billy Mitchell staged a bomber attack on an old battleship, sunk it, and claimed the battleship obsolete in the era of airpower. Japan began building a fleet of aircraft carriers in the 1920s, and the United States followed suit, introducing in the 1920s aircraft with brakes on the landing gear and an arrestor hook to stop a plane just seconds after it landed on a carrier. The 1930s and 1940s were a period of expansion. In 1940 alone Congress authorized the navy to build ten new carriers and in 1941 two more. These would complement the *Lexington, Enterprise, Yorktown,* and *Hornet,* the nucleus of America's fleet in the 1930s.

Life aboard a carrier was Spartan. Fresh water was in shortage. The *Hornet* kept only one bucket of soapy water for laundry and bathing and a second bucket of rinse water as its daily ration. Sailors followed use of these buckets with five or ten seconds under the shower, not long enough to adjust the water temperature, let alone remove all suds. Sailors on the carriers ate fresh food in their first weeks at sea, and then powdered milk, powdered eggs, and canned fruit, vegetables, and meat. With these gone, sailors subsisted on ground-up beans. Smaller ships served Spam and artificial potato, though crews were never quite sure what the latter was made from.

At sea the norm was fourteen to sixteen hours' work a day, with the day beginning two hours before dawn. Most of the activity centered on the launching and landing of planes. The flight deck was the scene of orders barked out and men in motion, often at a dead run. Men pushed the planes to the side and rear of the deck to permit each plane to launch. To recover the planes, however, sailors had to clear the rear of the deck by pushing aircraft forward. When planes were short of fuel, a crew sometimes could not clear the deck fast enough to accommodate the scores of planes that needed to land quickly. In 1951, for example, the carrier *Valley Forge* lost twenty-three of its 130 aircraft to loss of fuel or crash landings on the deck. Crashes invariably killed or injured the pilot and grounds crew. A plane that missed its arrester hook plowed into a nylon barrier and if the net broke caused yet another crash.

The crew of a carrier had to be alert to danger. Unlike an armored battleship, an aircraft carrier had a thin flight deck that bombs could penetrate. Below deck were fuel and munitions for the planes that could catch fire or explode. The *Enterprise,* for example, carried more than 200,000 gallons of fuel. Under these circumstances, the crew had to be adept at fighting fire should it erupt during an attack. An attack imminent, sailors drained the lines of fuel and filled them with carbon dioxide, which they could then use against a fire. A carrier also had foam-generating hoses that operated even when the carrier had lost power. Also, destroyers could contribute extra fire hoses to aid the men fighting to extinguish fire on a carrier.

The dangers of serving aboard a carrier led one sailor early in his enlistment to search the *Enterprise* for a safe place to hide in an attack, only to conclude that every position was vulnerable. When the *Hornet* lost power after a torpedo strike, the crew, rather than panic, began to hand out the ice cream in storage before it melted. In other instances an attack incited panic. When a bomb hit the destroyer *Smith,* two sailors, fearing that fire would engulf them, jumped overboard only to be pulled under by the sea. In the worst case a sinking ship spilled both oil and its crew into the sea. At times the oil caught fire, burning the survivors as they struggled to stay afloat. In other cases men could not avoid swallowing some of the oil along with seawater and became ill.

Technology affected daily life in the navy by forcing sailors to be ever ready for battle. In the days of sail, ships might not fight for weeks or even months. In the era of the aircraft carrier and the airplane, sailors in the fray could expect daily attack. Technology also made war more personal. The old ideal of battleships lining up to shoot one another from a distance with their guns was obsolete in the era of the carrier. Sailors saw bombers dive toward them and saw the flash of their machine guns and the release of bombs. When the Japanese turned kamikaze, they converted the airplane into a guided bomb. American crews now knew that the plane diving toward them would hit the deck and perhaps penetrate it. This sort of warfare was as close to hand-to-hand combat as the aircraft carrier and the airplane allowed.

THE SUBMARINE AND LIFE IN THE NAVY

Life aboard an American submarine was pleasant in at least one respect: as early as the 1930s it had air conditioning, not only keeping sailors cool but also giving them condensate with which to launder clothes, though a little condensate did not go far. Before the 1950s, American submarines ran on a diesel engine and a generator to charge the battery, which powered the lights, sonar, and computers. The essence of a confined space, a submarine had no way to dissipate odors and so stank of diesel fuel, body odor, and cigarette smoke. The carbon monoxide in cigarette smoke caused lethargy and headache, but it was not until 1990 that the navy commissioned its first submarine that forbade smoking.

In 1955 the navy launched *Nautilus,* the first nuclear-powered submarine, and by 1960 the navy had thirteen nuclear submarines. The fission of radioactive elements provided a nuclear submarine with electricity. These new submarines carried the Polaris, a missile that could launch a nuclear warhead 1,500 miles. Able to exceed twenty knots underwater, a nuclear submarine was five times faster than a diesel submarine. Officers and crew alike, however, fretted over the danger of an accident to the reactor and the spread of radiation. Even a single accident would undermine public confidence in nuclear power. Consequently, sailors spent long hours in study and paperwork. These activities, coupled with watches and cleaning, meant that sailors only had about five or six hours of sleep a night. The routine of sixty days at sea and thirty days in port for training also left sailors little time to observe birthdays, anniversaries, graduations, and the birth of children. Cut off from their families sailors could not transmit a message while at sea and could only receive messages no longer than forty words.

The submarines of the late 1970s had soft-drink machines and, a ubiquitous technology in the navy, ice cream makers, which American ships have had since 1905. By the 1990s submarines and ships in general had e-mail, satellite television, and gymnasiums. E-mail did much to break down a sailor's sense of isolation from his family. (Women may serve aboard a ship but only for days at a time. For example, female midshipmen may serve aboard a submarine but for only one night as part of their training.) In addition to providing e-mail, computers eased the work of logistics. Much as a clerk did at an auto-supply store, a sailor tracked parts and ordered replacements. Computers also allowed commanders to play war games in virtual space.

THE ATOMIC BOMB

Albert Einstein's equation $E=mc^2$ implied that a small amount of matter could yield enormous energy. In 1938 two German scientists, in bombarding uranium atoms with neutrons, split the nuclei of several atoms, releasing energy in accord with Einstein's equation. Thus the possibility arose that Germany might use uranium to build a fission bomb. Physicists Enrico Fermi and Leo Szilard, both European émigrés, believed that the United States should work to beat the Nazis to the bomb. Szilard convinced Einstein in 1939 to write President Franklin D. Roosevelt about the Nazi threat. Roosevelt established a committee but did not give the matter urgent attention. America's entry into World War II finally convinced the U.S. Army, which had oversight of the project, to pour scientists and money into what was code-named the Manhattan Project. In 1943 Fermi controlled the chain reaction of uranium atoms, and the enrichment of uranium and plutonium gave the army two uranium bombs and one plutonium bomb. In July 1945 the scientists of the Manhattan Project tested one uranium bomb in Alamogordo, New Mexico; that August, the Army Air Corps dropped the other two on Hiroshima and Nagasaki, Japan, killing more than 100,00 people.

Americans, both military and civilian, celebrated Japan's surrender on August 10 and credited the atomic bomb with ending the war. Many soldiers believed that they would otherwise have had to invade Japan, a battle that might have cost the United States one million causalities. From this perspective the atomic bomb seemed to be a miracle of science and technology, presaging a future of peace and atomic energy. Peace proved illusory, however. American soldiers were at war in the second half of the 20th century in Korea, Vietnam, and Iraq.

The use of the atomic bomb in World War II crystallized a new relationship between technology and war. Technology had for a long time made war more lethal, but now technology, in the form of nuclear weapons, could destroy humanity. Once the Soviet Union developed the atomic bomb in 1949, cultural commentators and some scientists feared that the next war would precipitate a nuclear exchange between the United States and the U.S.S.R., killing millions and perhaps exterminating humanity. The hydrogen bomb, far more destructive than the uranium and plutonium bombs, heightened the fear of extermination. Like the god of the Old

Testament, humans could destroy cities and all their inhabitants. Perhaps instead of saving humans, technology condemned them to death. Playing on these fears, filmmaker Sidney Lumet in 1964 released *Fail Safe*, based on the 1962 novel of the same name, in which a sequence of computer errors led to nuclear war. Humans had ceded control of technology to computers, with disastrous results.

In the nuclear age the ordinary soldier ceased to matter. In 1969 army general William Westmoreland lauded the use of technology that "will replace whenever possible the man with the machine."[1] The high death toll in Vietnam may have prompted Westmoreland's remarks, for Americans cared less about the destruction of machines than the loss of life. Whatever his motivations, Westmoreland surely appreciated the tendency of technology to usurp the role of soldiers. In the 20th century, particularly after World War II, technology could kill the enemy a continent away, with the soldier doing nothing more than pushing the button to launch a missile. The American soldier had become an extension of technology, the cipher who set the machine in motion.

VIETNAM, TECHNOLOGY, AND LIFE IN THE MILITARY

Even today, so much about the Vietnam War is murky. To the extent that the war was fought with technology, the Americans confused technology with victory. The technology that mattered, the technology that saved lives, was found in the Jolly Green Giants (a type of helicopter used to extract the wounded when enemy fire was too intense to permit other types of helicopters from doing the evacuation)] and the field hospitals. The Americans who never understood the war tried to use technology to bomb North Vietnam back to the Stone Age. The Americans who wielded this technology seldom questioned whether it could win the war. But technology had no way of contending with the jungle and with an enemy who subsisted on a handful of rice a day and who owned nothing more high tech than a bicycle. American technology could not alter the fact that the soldiers on the ground fought the sun, the heat, the rain, the jungle, and the enemy in short, ferocious firefights. Body counts, a low-tech way of keeping score, were the arbiter of victory.

By 1966 Secretary of Defense Robert McNamara had grown disillusioned with the strategy of heavy strategic bombing, which was a remnant of World War II. McNamara and his generals needed something high tech in Vietnam. They saw the war as a problem of supply: cut the supply lines along the Ho Chi Minh Trail and the North Vietnamese and Vietcong would have nothing with which to wage war. The use of troops to sever the supply lines was a low-tech solution. McNamara and the generals instead created a system of sensors, and between 1966 and 1972 the air force dropped some 20,000 electronic sensors on the trail. Many were camouflaged to look like plants. Some were even modeled to look like dog droppings. The realization that there were no dogs on the trail led to the redesigning of the sensors to look like pieces of wood. The sensors detected sound, motion, odor, even chemicals. When they were set off, the sensors emitted a signal to air force planes flying above the trail. The planes relayed the signal to Igloo White Infiltration Surveillance Center in Thailand, from which airmen radioed bombers with the location of the signal. On-board computers guided the bombers, and airmen at the center released the bombs once the planes were above the targets. In these instances the planes virtually flew themselves: technology had made pilots almost superfluous. Their daily lives were very different from the experiences of the grunts. Close air support was another matter. The jets, descending toward the ground, encountered antiaircraft fire and so pumped their pilots full of adrenaline. There was nothing routine about these missions.

The soldiers and marines who fought the war on the ground took little comfort in technology. For them the tropical climate pushed heat and humidity to levels they had never experienced in the states. Razor-sharp foliage cut them when they tried to hack their way through the jungle. During monsoons, the rain and mud made it impossible for them to stay dry while on patrol. Insects swarmed them and leeches sucked their blood. Technology did not mitigate these evils.

The North Vietnamese and Vietcong understood that the Americans were using electronic sensors against them. They fought back, hanging bags of urine, which the sensors could detect, from trees and played tape recordings of trucks. This deception, they knew, would bring American bombers. And so the North Vietnamese and Vietcong manned their antiaircraft guns, intent on shooting down the bombers.

VIDEO GAMES AND LIFE IN THE MILITARY

Conscription may have worked in World War II, but it failed in Vietnam, where morale plummeted as the war ground on to no discernable conclusion, leading Congress to end it in 1973. Thereafter, the armed services had to fill their ranks entirely with volunteers. The search for recruits led the military to the video arcades. The technology of video games demanded the hand-eye coordination that the military prized. In addition, many of the games were in quasi-military settings, giving a teen ample opportunity to kill his on-screen foes. These games rewarded aggression and violence, if not in real life at least in virtual reality. "Video games are really no different than war," said Chief Petty Officer Julia Reed.[2] No less enthusiastic about the militaristic virtues of video games was President Ronald Reagan. "Watch a 12-year-old take evasive action and score multiple hits while playing Space Invaders, and you will appreciate the skills of tomorrow's [air force] pilot," he enthused in 1983.[3] By then, military recruiters visited arcades in hopes of making an impression on the teens who excelled at the games. Once she had marked the best players, Reed played a game with them, offered to buy them a Coke, and filled their heads with the opportunities for advancement in the navy. "All of the people in those arcades are volunteers," said army general Donn A. Starry. "In fact...two-thirds of these games are military in nature—aircraft versus air defense, tank against antitank and so forth."[4]

The armed services could do more than find recruits in the arcades; they crafted their own video games to attract teens. The army used the video game *America's Army* to impress upon gamers that the army was "wicked and cool."[5] Aiming for realism, the game allowed players to fire only a fixed number of rounds before they needed to reload. Missiles traveled in arcs as they did in the real world. Players had to follow the army's rules of engagement. If a player shot a compatriot, the game ended with him in jail. *America's Army* was available at goarmy.com as a free download or from a recruiter as a CD. In its first two months, the game was downloaded 2.5 million times. Some of the players were not even Americans—Europeans, too, downloaded and played the game. In targeting teens, *America's Army* bypassed their parents, who were from the Vietnam generation and might have had reservations about a game that glamorized war. But soldiers also played the game. A soldier who entered his or her military e-mail address had the distinction of seeing his on-screen double wear the army's yellow star logo, announcing that the virtual soldier corresponded to a real GI. The army combined the game with an effort to recruit African Americans, who make up a disproportionate share of recruits. So successful was *America's Army* that the army released a companion game, *America's Army: Special Forces.* The game included patriotic music and video clips of real Special Forces, with the hope of increasing the number of recruits to them.

So enamored were the services with video games that they crafted their own games as training devices. In 1977 the Army Armor School modified the commercial game *Panther PLATO* into *Panzer Plato* for training tank crews. Other games followed. *Army Battlezone,* a product of the cold war, pitted the Bradley Armored Vehicle against Soviet tanks and helicopters. The game also included NATO tanks, putting a premium on the quick differentiation between enemy and allied tank. Soldiers operated the game through a steering wheel and switches that simulated the control panel of the Bradley. The aim was to simulate combat as well as provide soldiers the vicarious enjoyment of killing their virtual foes. In 1997 two marine lieutenants modified the commercial game *Doom II* to create *Marine Doom.* They replaced aliens and demons with opposing forces, scans of GI Joe figures. *Marine Doom* armed a marine with the standard issue: an M16 rifle, an M249 automatic weapon, and an M67 grenade. The creators of *Marine Doom* conceived of it as a technology to strengthen cooperation among the four men of a marine squad. The four decided how best to set up fields of fire, how best to outflank the enemy, and how best to pin them down with automatic fire, thereby allowing the other marines to race ahead and attack them in the rear. Marine Corps commandant General Charles Krulak recommended that Marines play *Marine Doom* when under stress, since combat was inherently stressful. Krulak favored the playing of *Marine Doom* after a grueling physical training session or march or when deprived of sleep.

One use of technology promises to restrain the use of armaments. Computers can simulate war without soldiers suffering casualties. In this sense war becomes a virtual exercise. The rise of virtual war paralleled the rise of video games that likewise simulated war. Virtual war and video games seek to reduce violence to a cerebral rather than a real activity.

Military Technology in Peacetime

In the 20th century the technology of war became too lethal to use. As such, it occupied a unique position in the United States. Americans prize mainstream technology—vacuum sweepers, washers and dryers, air conditioners, and furnaces—because they can use these devices to better their lives. In contrast, military technology has the potential to destroy incalculable numbers of people. Rather than use this technology, Americans prefer peace. We live with this technology as a necessary evil.

NOTES

1. Ed Halter, *From Sun Tzu to Xbox: War and Video Games* (New York: Thunder's Mouth Press, 2006), 104.

2. Ibid., 142.

3. Ibid., 117–18.

4. Ibid., 138.

5. Ibid., ix.

8

EDUCATION

As the United States came to rely on scientific expertise and technical knowledge, the need to educate children in these subjects became important. Whereas the United States had only 100 public high schools in 1860, it had more than 12,000 in 1914. By 1900 thirty-one states required school attendance of their children. Notwithstanding this progress, schools had much to do. Those in the countryside were underfunded, and some southern blacks had no schools to attend. These disparities penalized minorities and the poor.

Science was part of the curriculum at most schools from an early date. Reformers in the late 19th century, aware that the United States had come to rely on scientific knowledge, urged schools to add more science to their course offerings. Much effort focused on the elementary schools, with the rationale that teachers could impress the habits of observation, classification, and other methods of science onto malleable young children. Building on the elementary school curriculum, secondary schools added chemistry, physics, and the new biology to their curriculum. Laboratory exercises constituted a large part of a course—students in the early years of the 20th century actually spent more time in the laboratory than did students at century's end. Laboratory exercises sharpened skills in observation, classification, and, under the best circumstances, stimulated pupils to think like scientists. Aside from the laboratory, students spent much class time reading their textbook, with the teacher serving as the lead reader and calling on students to read passages from the text. As an exercise in reading, this activity many have succeeded, but as a stimulus to curiosity in science it must have failed. Circumstances were not dire everywhere, though. Results varied from school to school, and much depended on the quality and enthusiasm of the teacher.

THE COMMITTEE OF TEN AND SCIENCE EDUCATION AT THE DAWN OF THE 20TH CENTURY

In July 1892 the National Education Association, meeting in Sarasota Springs, New York, appointed a committee to order and reduce the number of science courses offered by high schools, with the aim of keeping in the curriculum only those courses that colleges and universities required for admission. At the same time, the NEA hoped that colleges and universities would standardize their admission requirements in response to the reforms at the high school level. Charles Eliot, president of Harvard University, headed the committee, and he recruited the U.S. commissioner of education, the presidents of the University of Michigan, the University of Missouri, the University of Colorado, Vassar College, and Oberlin College, and three high school principals. With its ten members, the committee was known as the Committee of Ten. Ambitious from the outset, the committee resolved that science courses should make up one-quarter of secondary school offerings. To determine exactly what should be taught, the committee appointed subcommittees: the Conference on Physics, Chemistry, and Astronomy; the Conference on Natural History; and the Conference on Geography.

THE CONFERENCE ON PHYSICS, CHEMISTRY, AND ASTRONOMY

The Conference on Physics, Chemistry, and Astronomy urged that high schools require students to study physics and chemistry but not astronomy, which might remain an elective or be jettisoned from the curriculum. Instruction in physics and chemistry, it was decided, should begin in the elementary school and be taught by experiments rather than lecture and textbook readings. Chemistry should precede physics in the curriculum to help students gain competence in the mathematics that they would need for physics. Each course should be taught for one year, with half the class time being devoted to laboratory exercises. Laboratories should teach students to measure and calculate the properties of objects in motion and of natural phenomena. Standard exercises were the calculation of an object's density, the tracing of the lines of force of a magnet, the measurement of the resistance of a wire as electricity moves through it, the calculation of the momentum of an object in motion, and the like. Laboratories should also reinforce lessons about the physical laws learned in class, such as about the freezing point of water.

THE CONFERENCE ON NATURAL HISTORY

Natural history was the 19th-century equivalent of biology. High schools did not teach a course entitled Natural History but rather three courses that in the 19th century were a part of natural history: physiology, botany, and zoology. The contents of physiology were a response to social conditions in 19th-century America. Reformers in the late 19th century grew alarmed at the filth and the spread of diseases in cities and so shaped high school physiology to emphasize the habits of hygiene and health.

The Conference on Natural History recommended that high schools offer physiology as a half-year course following chemistry and physics. Unlike the Conference on Physics, Chemistry, and Astronomy, the Conference on Natural History doubted the value of laboratory exercises and instead urged teachers to use a textbook. Of the other two natural-history courses, the conference recommended that high schools require only botany or zoology but not both, leaving the other as an elective or omitting it from the curriculum. Botany or zoology should be, conferees believed, a full-year course with laboratory work three days a week. Because the focus was on the laboratory, teachers and students needed not a textbook but a laboratory manual. Eager to infuse the spirit and method of science into young children, the Conference on Natural History recommended that schools introduce life science in primary school. Teachers were to forgo a textbook and instead direct students to observe nature.

THE CONFERENCE ON GEOGRAPHY

The Committee of Ten included geography among the sciences because in the late 19th century geography had included geology and meteorology. The Conference on Geography recommended that schools teach geography as a survey to elementary school children, giving them a foundation for in-depth study in high school. In the grades leading up to high school, conferees believed, students should study physical geography, with high schools offering physiography, meteorology, and geology as half-year electives that students might take after completing chemistry and physics. To make the study of the earth concrete, conferees urged teachers to take students on field trips to illustrate some feature of earth's surface. In contrast to the Conference on Natural History, the Conference of Geography recommended the use of a textbook but warned against students' memorization of terms and definitions, based on the belief that rote learning was quickly lost after a test or quiz.

The Themes of the Conference Reports

Conferees in nearly every discipline of science deemphasized textbooks in the belief that teachers who taught by textbook fell back on rote learning. Rote learning, in the view of the conferees, was not learning at all. Rather than trudge through 400 pages of text, students should learn science by doing experiments and observing nature. This empirical approach encouraged students to amass data, which the teacher would help them form into a generalization. Science was therefore to be Baconian in approach.

TECHNOLOGY IN THE CLASSROOM

Cinema captured the attention of educators early in the 20th century. Aware that motion pictures enthralled Americans, teachers hoped they would stimulate enthusiasm for science in their students. Educators believed cinema to be "the next best thing" to direct observation of nature. The ability of film to depict what students could not observe led teachers to select films on aspects of nature that were too large or too small to be captured by the senses: the solar system, bacteria, the structure of the atom, and the like. During the 1920s, Kodak produced some 250 films for the science classroom, and by the early 1930s motion pictures had become standard in the science curriculum. In the 1930s films with a voice narrative replaced silent films. In the 1960s teachers began to use short films that came in a cartridge and did not need to be threaded into the projector. These short films illustrated a single concept in keeping with the pedagogical principle that students learned best when teachers taught them one new concept each lesson. As early as the 1950s, video emerged to challenge motion pictures, but not until the 1980s did teachers begin to replace their projectors and films with video and television sets. Video gave teachers the ability to record television programs for later viewing in class and for students to make their own videos of a Venus flytrap capturing an insect, of sulfuric acid burning a hole in a piece of paper, and so forth.

Television complemented motion pictures in the science classroom. Teachers could tape a science program for later viewing, or their class could watch a program at the time of broadcast, if the program aired during school hours. The latter did not allow teachers to preview content, however, and to prepare an accompanying lesson in detail. Despite this shortcoming, science television enjoyed popularity in the 1950s. Iowa State University, for example, ran television station WOI-TV. Between 1952 and the mid-1970s, WOI-TV broadcast *TV-Schooltime* at 10 A.M., Monday through Friday, in thirty-minute segments. Along with the program, WOI-TV issued a teacher's guide that suggested corresponding activities for students to do before and after the program. The content was not adventurous. One program, for example, showed how a teacher and her students set up an aquarium. In 1955 WOI-TV expanded its science offerings with *Chemistry 101*, a program to complement instruction in an introductory high school course in chemistry. Against the charge that science television was boring, Bill Nye offered the MTV generation a program that combined quick cuts between scenes and music that spoofed the videos popular on MTV.

Perhaps the most promising technologies of the late 20th century were the computer and the Internet. President Bill Clinton championed the ideal of computers in every U.S. classroom, though at the end of the 20th century disparities remained between rich and poor districts and between suburban and both urban and rural schools. Schools with computers and Internet access set up their own Web pages with links to science content, such as virtual tours of NASA or the Smithsonian Institution. As part of a lesson on the planets, students in one school interviewed a NASA scientist in real time and e-mailed students in another state who were studying the same science concepts. Parents e-mailed teachers to learn of their child's progress in science as well as in other subjects. Teachers posted activities, reading lists, and homework online for students and parents to consult. By the end of the 20th century, many elementary students could not recall a time when they did not supplement classroom instruction with online content.

NATURE STUDY

The Committee of Ten recommended that students in grades one through eight study nature one hour a week. Again textbooks were eschewed, and the focus was to be, in the words of Harvard zoologist Louis Agassiz, "nature, not books." Better to lead students outdoors in fresh air, where they could observe a spider weaving a web or collect leaves and rocks, than for teachers to drill students on information memorized from a book. An architect of the movement to teach the study of nature in the elementary schools was Cornell University's Liberty Hyde Bailey. An agricultural scientist by training, and dean of the College of Agriculture, Bailey used the university to promote nature study. Cornell University distributed free literature on nature and agriculture to 25,000 of New York's 29,000 certified teachers. Bailey hoped that instruction in nature science would give students an appreciation of rural life. Nature study taught elementary students to collect insects and other specimens and to make careful observations, usually pencil-and-paper drawings, of these specimens. The use of these data to reach a generalization was beyond the capabilities of elementary school children—at least, according to the pedagogy of the time—and so was left to high school students. This study of nature countered the influence of urbanization on America's children. The city, full of concrete and glass, isolated children from nature. In 1883 psychologist G. Stanley Hall reported that children had so lost touch with nature that they could not identify trees or a field of wheat. Some did not even know the seasons. The nature-study movement drew strength from the conservation movement in its desire to preserve the pristine beauty of nature, and both movements saw in nature an antidote to the ills of life in modern America. Nature study also drew strength from natural theology, which purported that to study nature was to study the handiwork of God.

From its inception in the 19th century, nature study grew so popular that by 1925 virtually every school taught it. Under the encouragement of superintendent Francis Parker, students at the public schools in Quincy, Massachusetts, brought insects and amphibians to class and took nature hikes through fields and along streams. They recorded their observations in a notebook, made drawings of specimens, performed calculations, and planted gardens to study plant growth. Students compared the habitat of various species, tested the effect of different soils on plant growth, observed the life cycle of butterflies, and recorded the weather. Fifth graders in California's Oakland City School District studied the growth of the pistil in flowers, the anatomy of earthworms, snails, and slugs, and the properties of magnets.

Nature study was overwhelmingly biological in content. Historian Kim Tolley notes that nature study, in attracting girls, reinforced their pursuing careers in botany and nursing. In 1924 fourth grade girls at a New York elementary school studied ants, collected fossils, and reported on reptiles. First grade boys, on the other hand, eschewed the emphasis on the life sciences but wired a dollhouse with lights and a doorbell.

Although some schools retained nature study into the 1940s, the movement began to lose momentum in the 1930s. The spontaneous, eclectic

This girl appears to be doing a titration. Some schools allotted two periods to science, one for the traditional lecture and the other for laboratory investigations. Courtesy: Shutterstock.

character of nature study, once a virtue, struck educators in the 1930s as unsystematic. Teachers reintroduced textbooks, the antithesis of the spirit of nature study, to elementary science in hopes of systematizing what their students learned. Once distinctive in the freedom it afforded students, elementary science became just one of many subjects that relied on textbooks, memorization, and tests.

THE RISE OF BIOLOGY

The biological content of nature study prepared students to take courses in the life sciences in high school. Around the beginning of the 20th century, students in most schools selected among the familiar triad of physiology, zoology, and botany, all electives. In 1910 15.8 percent of high school students studied physiology; 7.8 percent, zoology; and 16.3 percent, botany. There was in most schools no single life-science course that unified the study of plants and animals. Biology emerged in the early 20th century as that science class. In the 1890s schools in Chicago and in 1902 Dewitt Clinton High School in New York City began to offer biology as a full-year course instead of physiology, zoology, and botany. By 1928 13.3 percent of high school students took biology class, whereas 2.7 percent took physiology, 0.8 percent took zoology, and 1.6 percent took botany.

The first biology courses taught physiology, zoology, and botany as separate units. But by the 1920s biology had taken a practical orientation. Students studied the calories and protein in food in relation to its cost, the causes and prevention of food spoilage, the dangers of alcohol and illicit drugs, the advantages of sleeping on tenement balconies in hot weather, the role of vitamins in nutrition, the value of exercise and good posture, and the benefits of open bedroom widows for ventilation, dust-free air in factories and the home, sanitary water and a sewage system, clean eating utensils, garbage disposal, and quarantining the sick. Students also learned to identify insect pests and to destroy their habitats.

A staple in life-science courses in 1900, dissection fell out of favor in the 1910s. Some students and parents thought the practice of cutting up a frog, a cat, or some other animal repugnant. The use of cats in dissection led to the charge that unscrupulous people were stealing cats to supply the schools, and newspaper accounts heightened that fear. Some school administrators bought embalmed cats to avoid the tint of scandal. Others, averse to criticism, eliminated dissection and the bad publicity that came with it.

REFORM IN BIOLOGY EDUCATION

The practical orientation of high school biology courses left little room for the study of evolution, which most textbooks, when they mentioned it at all, classified as a theory. Elementary school nature study and high school biology had introduced students to the diversity of life, but these subjects had done little to help students make sense of this diversity. The obvious answer was to introduce students to idea of the evolution of life by natural selection. Few schools and textbooks made much of an attempt to teach evolution even before the furor of the Scopes Trial in 1925. One textbook devoted only eight of its more than 400 pages to evolution, an amount a teacher could safely ignore without compromising the content. The argument for including evolution in the biology curriculum centered on the principle of common descent: that all life, no matter how different any two kinds are, is related through a common ancestor. One might think of the first life as the alpha ancestor whose descendents diverged over millions of years into the current diversity of life. The rich diversity of life was not just "sound and fury signifying nothing," it was the outcome of millions of years of evolution. Evolution itself was not some random force but operated according to the mechanism of natural selection.

Those schools that taught evolution in the early 20th century irked people who believed in a literal interpretation of the Genesis story, which implied that God had created all life in its present form some 6,000 years ago. The most vocal of the creationists was William Jennings Bryan, a skilful orator who had thrice been the Democratic nominee for president. Bryan objected to the premise that educators could use their professional judgment in deciding what to teach. Rather, teachers were "hired servants" beholden to the taxpayers who paid their salaries. Because taxpayers in America were overwhelmingly Christian, according to Bryan, science instruction must conform to Christian doctrine. In contrast to the mass of Christians stood what Bryan characterized as a small elite of atheists, skeptics, and scientists who had forced evolution on the schools in defiance of the Christian majority.

In several states the creationists imposed their will. In Portland, Oregon, fear of controversy was enough to prompt the school superintendent to direct teachers in the 1920s to mention neither evolution nor creationism in the biology classroom. In 1923 the Oklahoma legislature banned the use of textbooks that included evolution among their topics. In 1925 North Carolina's Mecklenburg County Board of Education enacted a resolution against the teaching of evolution in its schools and for the removal of books that mentioned evolution from school libraries. That year, Texas governor Miriam Ferguson banned from public schools textbooks that mentioned evolution, Tennessee banned the teaching of evolution in its public schools, precipitating the famous Scopes Monkey Trial, and the Mississippi state superintendent instructed his teachers to omit evolution from their biology lessons. In 1926 Mississippi, as had Tennessee, prohibited the teaching of evolution in its public schools. That year Atlanta, Georgia, banned the teaching of evolution in its city schools. In 1927 local school boards in Indiana and Kansas decided against hiring teachers who believed in evolution and for the removal from school libraries of books on evolution. In 1928 voters in Alabama passed an initiative to prohibit the teaching of evolution in the state's public schools. Many teachers and textbook publishers played it safe by omitting to the topic. For example, the textbook *Biology for Beginners* by Truman Moon put a picture of Charles Darwin on the frontispiece of its 1921 edition. By 1926, however, the publisher had deleted the picture of Darwin as well as any mention of him. Instead, the frontispiece was an illustration of the human digestive tract.

The launch of *Sputnik* in 1957 put the creationists in momentary retreat. Sputnik had demonstrated, many scientists and policymakers believed, that Soviet science and technology had overtaken science and technology in the United States. Scientists blamed the poor quality of science instruction for allowing the Soviets to take the lead. The federal government thus committed money to the improvement of science instruction, giving physics priority, though biology claimed its share of the federal largesse. In 1958 the National Science Foundation earmarked $143,000 to establish the Biological Sciences Curriculum Study, and the scientists in this committee drafted three biology textbooks, all of which covered evolution. By the mid-1960s nearly half of America's schools used one of these texts.

Students may not have given much thought to evolution in their daily lives, but the reform of biology study in the 1950s brought the issue to their attention. The first reform integrated evolution and genetics. Like evolution, genetics had received little space in the biology curriculum in the first half of the 20th century, despite the fact that the United States was in the forefront of genetics research. This lack of attention made it too easy for students to misunderstand what genes were and how they expressed themselves in an organism. Some students thought that an organism with a gene for a trait always expressed that trait. That is, students did not distinguish between genotype and phenotype. Other students were not sure what a mutation was but thought it was common and always harmful. Still others believed genetic diseases spread like contagion and that other diseases were predetermined by race.

As in physics, reform came from above. Between 1949 and 1953, geneticist Anthony Allison demonstrated that the disease sickle-cell anemia confers partial resistance against malaria. Sickle-cell anemia is a recessive trait, and a person who is homozygous for the gene that expresses sickle-cell anemia experiences the worst symptoms. One might expect that natural selection would eliminate the gene for sickle-cell anemia from the population. Simply put, those who suffer from the disease should have on average fewer children than the healthy. This differential success in reproduction should over time delete the gene from the human genome. Allison demonstrated, however, that people who are heterozygous for the sickle-cell gene have a mild form of the disease and are partially resistant to malaria. For this reason the sickle-cell gene remains in the population despite what would otherwise be intense pressure from natural selection to delete it. High school students who studied sickle-cell anemia in their biology class learned several interlocking concepts: natural selection, gene, genotype, phenotype, recessive gene, dominant gene heterozygosity, homozygous, and genetic disease.

The second reform came after lepidopterist H. B. D. Kettlewell in 1956 confirmed the phenomenon of industrial melanism. The organism under scrutiny was the peppered moth, an insect that comes in both dark and light hues. As early as the 19th century, observers noticed that dark-hued moths outnumbered their light-hued counterparts near cities. In the countryside, by contrast, light moths were more numerous than dark moths. Light moths, British biologist J. B. S. Haldane surmised in 1924, rested on trees covered by lichen that formed a light covering over the bark. The light color camouflaged light moths, and they escaped predatory birds that instead recognized and ate dark moths. Near cities the reverse applied. The soot from factories covered trees and killed the lichen, giving dark moths the advantage of camouflage. Birds then preyed on light moths. Kettlewell confirmed Haldane's insight by capturing, tagging, releasing, and recapturing moths of both hues with the expected results. Even more than the lesson on sickle-cell anemia, the lesson on industrial melanism became the standard introduction to natural selection. Evolution could at last claim a place in the biology curriculum.

Yet the commitment to evolution was not as deep as one might have hoped. A 1961 survey of 1,000 high school biology teachers counted two-thirds who believed they could adequately teach biology without teaching evolution. Sixty percent believed that evolution was nothing more than a theory. But if the support for evolution was lukewarm in some quarters, opposition was hardening. Embracing the idea that evolution was just a theory, creationists began in the 1960s to insist that other theories of human origins be taught. It was only fair, after all, to let students choose which among competing theories to believe. Creationists offered their own "theory" of human origins in the biblical account of creation. Bent on making creationism appear to be a science, creationists recruited scientists to their ranks and established institutes that purported to do research on human origins. They wrote their own textbooks, which they printed in-house and offered for sale to school districts. In the 1970s and 1980s several school districts mandated the teaching of the biblical story of creation alongside the teaching of evolution. In 1981 Arkansas and Louisiana enacted equal-time laws, but a federal court struck down the Arkansas statute and the U.S. Supreme Court voided the Louisiana law. Defeat in Arkansas and Louisiana did not discourage creationists, however, and they continued to shape the teaching of biology at the end of the 20th century.

THE RISE OF GENERAL SCIENCE

In the first years of the 20th century, the sciences appeared to be in retreat. Enrollments in physics, chemistry, astronomy, and physiology—indeed, in every science but the new biology—had fallen steadily since the 1890s. Reformers interpreted these declines as evidence that the science curriculum did not meet the needs or interests of students. From this perspective reformers in schools throughout the United States experimented with new courses. In 1903 the public schools in Springfield, Massachusetts, sought to pique students' interest with an introductory physics course, one shorn of the mathematics that turned off so many students. The public schools in Columbus, Ohio, taught a mix of chemistry, physics, and physiography in a single course, and the public schools in Oak Park, Illinois, offered a course that comprised physics, chemistry, and physiology. These latter courses functioned as a survey of the sciences, a measure to counter the belief that science had become too specialized and technical to appeal to students.

By getting away from specialization, these new courses could focus on teaching students to appreciate the value of science. In this context science was synonymous with technology, and educators, aware that the industrial revolution had sparked widespread interest in technology, were ready to blur the distinction between science and technology and to offer students technology as a substitute for science. The new general science course was really the study of the applications of technology to daily life. One general science textbook in 1912 taught students about heat and its use in cooking, photography, and the lightbulb; air and water pumps; the rolling of glass to make windows; the properties of soap; and the effect of headache powders. *The Elements of General Science* (1914) included the topics air pressure and temperature and their relation to furnaces and chimneys, water pressure and pumps, and methods for reducing bacteria and dust in the home. University High School in Chicago offered a general science course in which students studied electric generators, gasoline and steam engines, refrigerators, sewage systems, and methods for purifying water.

By 1930 schools in every state taught general science and had integrated it in the curriculum in the familiar sequence: general science, biology, chemistry, and physics. In 1934 800,000 students (18 percent) took a general science class. For many, general science was the only high school science course they took. In this case general science served, curiously, as both students' entry and exit from science education.

HIGH SCHOOL PHYSICS

Physics may have been the subject least influenced by the Committee of Ten. The recommendation that physics students spend half their time in class doing labs foundered against the reality that some high schools did not have or chose not to spend money on laboratories. In 1934 a University of Pittsburgh physicist reported his disappointment that so few high school in Pittsburgh had laboratories. Instead of lab work, students listened to lectures, watched demonstrations, and performed calculations. This approach bothered G. Stanley Hall, who as early as 1901 complained that high school physics was too quantitative. He believed that students wanted a "more dynamic physics." University of Chicago physicist Charles Mann agreed, citing the limited pedagogical value of quantification and measurement.

The implication was that physics, at least at the University of Chicago, provided a richer learning experience than the lectures and calculations high school students endured. The reality was that college physics could not claim a distinction, since it resembled high school physics. Teachers at both levels often used the same textbook because high schools aimed to prepare physics students for college. Not merely a science, physics was an elite discipline that brought students a step closer to college. With a reputation for abstractness and difficult mathematics, physics lost pupils in the 20th century as the majority of students looked for something easier or more relevant to study. Whereas 22 percent of high school students had taken physics in 1890, the percentage slipped below 15 percent in 1910, 7 percent in 1928, and 5 percent in 1965.

The daily classroom routine of high school physics students was remarkably stagnant, changing little between 1930 and 1950. They studied mechanics, gravity, heat, sound, light, magnetism, and electricity. Writers had codified these topics in textbooks in the 19th century, and there they ossified during the first half of the 20th century. Not only was high school physics unresponsive to the recommendations of the Committee of Ten, it failed to absorb and to educate students in the revolutionary issues in physics in the early 20th century: atomic structure, quantum theory, and relativity. Students in the 1950s still learned nothing about the new physics, at least not unless they read outside of class.

Reform came slowly after World War II. The public perceived that physicists had played a prominent role in the Allied victory. Some went so far as to characterize World War II as the physicists' war. A poll in 1946 ranked physicists first in esteem among professionals. Returning to their universities after the war, physicists used their prestige to break high school physics out of its 19th century mold. The first efforts at reform were small. The 1952 textbook *Everyday Physics* took a practical approach, much as high school biology had in the 1920s. Students who read *Everyday Physics* studied chapters on washing machines, science and sports, and safe driving. More far-reaching were the reforms of the Physical Science Study Committee. In 1957 and 1958 a small number of schools implemented the PSSC reforms, which required students to study a small number of topics in depth. PSSC physics used laboratory work, as the older physics had not, to train students in the process of inquiry, in effect teaching them to model the thinking of physicists. Students responded to PSSC physics. The commotion in one high school boys' locker room brought the principal to investigate. He found several boys gathered in the shower, using their new stroboscopes to illuminate water droplets as they fell from the showerhead.

SCIENCE EDUCATION AT THE END OF THE 20TH CENTURY

Two events shaped science education at the end of the 20th century. The first was Rachel Carson's publication of *Silent Spring* in 1962. By underscoring the extent to which DDT had harmed wildlife, *Silent Spring* launched an environmental movement. The new environmental movement was sensitive to the potential of science to aggravate environmental problems. Some environmentalists distrusted science or limited the problems to which it might be applied. The second event was the Southern Corn Leaf Blight of 1970. That year farmers lost fifteen percent of their corn crop, and Americans learned, as the Irish had a century earlier, that genetic uniformity makes a crop vulnerable to an epidemic.

In addition to these two events, a third rumbling arose in the circles of education. Accusations that the curricula in schools were flabby led to calls for reform. Responding to political mandates, schools instituted tests that students needed to pass to graduate from high school. The science portion of these tests shaped what students learned in the late 20th century.

The environmental movement found its way into the science curriculum in a multitude of ways. Students studied the fragility of wetlands and the danger of pollution. Much as nature study had earlier in the century, elementary schools jettisoned their textbooks so that students could study life in its own environs. Students now observed the germination of a plant, the hatching of a chick, the life cycle of a frog, and the feeding habits of lizards. The lesson of genetic diversity led teachers and textbooks to emphasize the diversity of life and the genetic plasticity of species. Students in some schools went a step further in learning to sequence their DNA. The science tests pushed students to study the orbit of the earth around the sun, to appreciate that earth's tilt on its axis is responsible for the seasons, to use simple machines, and to identify the fulcrum in a lever. Science education at the end of the century emphasized basic skills rather than an advanced understanding of science.

In the 1980s a movement arose against the post-*Sputnik* reform of science education. The movement drew its dissatisfaction with science education from surveys of students. The longer students had been in school and the more science courses they took, the more they disliked science and the less relevant they believed science to be to their lives. Traditional science courses made students less rather than more curious about the natural world. Whereas science education in the 1960s and 1970s emphasized theory as the intellectual structure that linked facts with one another, the Science, Technology and Society (STS) movement deemphasized the value of theory. Rather, proponents claimed that science education should, as it did in the early 20th century, have a practical approach and that knowledge for its own sake was a quaint but erroneous perspective. To have value, knowledge had to solve some problem. The STS movement valued science education that was relevant to the lives of students. Ninth graders in one science class performed chemical analyses of lipstick, prepackaged chocolate chip cookies and other items to determine the amount and kind of preservatives in them. Other students compared the strength of paper towels and the sogginess of breakfast cereals and milk, the amount of residue left on the hands after washing with different brands of soap, and the strength of different brands of nail polish.

From the outset the STS movement drew connections between science, technology, and everyday life, with the goal of clarifying the ways in which science and technology shaped everyday experiences. This approach concentrated on problems: the depletion of fossil fuels, the Malthusian crisis, the disposal of nuclear waste, and the like. Students were to engineer their own solutions to these problems rather than to await a pronouncement from their teachers. Students were even to design, if unsystematically, what they were to learn. In one example of this, a teacher scuttled a presentation of careers in chemical engineering when he perceived the students' attention shift from the content of the presentation to the thunderstorm that was raging outside the classroom. When a student wanted to know the amount and kind of acid in rainwater, the teacher made this question the focus of the class. The students collected rainwater and determined its acid content. They asked students in other schools to do the same and to share their results. They telephoned scientists in Washington, D.C., and an ecologist at the state department of game and fish to learn of the dangers of acid rain. They made pamphlets on the effects of acid rain on the environment and distributed the pamphlet at a meeting of the city council. In another science class, interest focused on finding the cause of the dinosaurs' extinction. All but three of the students adopted the now-conventional explanation that the impact of a meteor, and the harm it did the biosphere, caused the dinosaurs to die off. The three holdouts, however, thought disease the culprit and were elated some weeks later when a journal article articulated this hypothesis.

In taking direction from students, the STS movement sought to emphasize depth of inquiry rather than breadth of coverage. At one school, students spent an entire year on studying the harmful effects of ozone depletion. The STS movement also devalued textbooks. They were to be a resource that students consulted only when they deemed it necessary, rather than being the core book that students used everyday. Textbooks were no longer to be authoritative but rather were "to contain errors and information of dubious quality" in the name of teaching students to distinguish truth from error.[1] Inaccuracies ran deeper than the textbook. One teacher never corrected students for offering an incorrect answer to a question.

The STS movement encouraged a teacher to raise "issues and questions even if you know them to be false or irreleavant."[2]

At the end of the 20th century education offered students new insights into the sciences. It had long been possible for students to type their blood. Now they could sequence the DNA in a drop of blood or in tissue extracted from their mouth with a Q-tip. These are laboratory activities, and if the trend toward learning by doing continues, the laboratory may become as important as it was early in the 20th century.

Curiously, the textbook remains a fixture in science class. In some schools little seems to have changed since early in the century. Teachers still function as lead readers and still call upon students to read passages of text. Perhaps educators value rote learning, but science is more than rote learning. It is a way of thinking not a of memorizing, and this realization should influence the way students learn science.

At its best, science challenges students to observe the world carefully, to generate a hypothesis to explain some behavior in the world, and to devise a way to test the hypothesis. This reasoning has been at the heart of science for centuries and the schools that instill it in their students have made them capable amateur scientists. Many schools aim to produce students who think like scientists. State proficiency exams test students' knowledge of science in its particulars and in its methods. States grade schools on how many students pass the exam, leading to invidious comparisons among schools. For many students science is not a discipline to be learned but a test to take. Fear of failing the test rather than curiosity about the natural world motivates many students to study science. The test taking mentality of students implicitly denies the worth of science. Science is simply one more swamp to be trudged through. But taught by capable, enthusiastic teachers, science has the potential to woo students to study the natural world with the curiosity that enlivens the experience of learning.

NOTES

1. John Lochhead and Robert E. Yager, "Is Science Sinking in a Sea of Knowledge? A Theory of Conceptual Drift," in *Science/Technology/Society as Reform in Science Education,* ed. Robert E. Yager (Albany: State University of New York Press, 1996), 34.

2. John E. Penick and Ronald J. Bonnstetter, "Different Goals, Different Strategies: STS Teachers Must Reflect Them," in Yager, *Science/Technology/Society as Reform in Science Education*, 171.

9

LEISURE

At first the concept of leisure was at odds with American capitalism. Andrew Carnegie closed his steel mills only one day a year, July 4. Many businesses demanded a six-day workweek, leaving only one day for leisure. Americans made the most of this one day. Some of them took to the roads on a bicycle, a machine that wedded technology and leisure. Young women as well as men rode bicycles, finding in them freedom from the watchful eye of parents. Women took them on streetcars, shopping by bicycle in the city.

Leisure came into its own after World War II. The postwar affluence gave middle-class Americans in suburbia the time and money to indulge their proclivities for leisure activities. The bicycle had by then been transformed into a sleek and aerodynamic ten-speed. Surfing, an ancient leisure activity, grew in popularity in the 1950s and 1960s. Surfboards shrank from the twenty-foot-long olo boards of Hawaii to just six feet by the 1970s. Polystyrene and fiberglass replaced wood in surfboards. Kids in suburban California invented the skateboard and with it developed their own subculture. Computer scientists invented video games, some of which evolved a level of violence that prompted Congressional scrutiny. Leisure was not exclusive to affluent white males. Women and minorities also enjoyed leisure activities .

THE BICYCLE

Although Leonardo da Vinci sketched a bicycle with handlebars, pedals, a chain drive, a seat, a frame of sorts, and wheels with spokes, the

In the first years of the 20th century the bicycle gave men and women freedom of movement. For most Americans the automobile replaced the bicycle though the latter remained popular as a leisure pursuit. Courtesy: Library of Congress.

drawing was not known about until 1966 and so did not influence the design of the bicycle. In 1791 French nobleman Comte de Sivrac displayed the first bicycle, a wooden frame with two wheels but no pedals. A rider moved the bicycle, known as a velocipede, by pushing his feet against the ground and coasting, much as skateboarders do today. Around the mid-19th century, three men independently claimed to have invented pedals. Attached to the center of the front wheel, the pedals gave a cyclist greater speed than had been possible by pushing against the ground and coasting. The larger the front wheel, the faster a cyclist could go, and in the 1870s manufacturers built bicycles with enormous front wheels. A cyclist, seated just behind the center of gravity, was more than five feet above the ground. The tendency of this type of bicycle, known as an ordinary, to tip forward imperiled a cyclist. Departing from the hazards of an ordinary, John Starley of England designed the safety in 1885. The safety had two wheels of equal size, a diamond-shaped frame, pedals attached to the frame rather than to the front wheel, a chain drive, and coaster brakes: all features of some modern bicycles. In 1888 John Dunlop replaced the solid rubber tire with the pneumatic tire for greater shock absorption and lighter weight.

At the dawn of the 20th century Americans adopted the bicycle for transportation and leisure. As an inexpensive technology whose price had fallen from $400 in the 1880s to $20 in 1898, the bicycle democratized leisure by appealing to both the middle and working classes and to both men and women, and in doing so shaped the daily lives of Americans. Men took rides through the countryside on Sunday, often their only day off during the week, and women, despite the uneasiness of some men, took up cycling to challenge Victorian notions of what constituted leisure for women. Corsets and ankle-length skirts would not do; women instead bared their legs in short skirts and bloomers. The straddling of a bicycle struck some men as too provocative a position for women. No less threatening were young women who took up cycling to court men beyond the supervision of parents. This use of cycling as a means of courtship prefigured the enthusiasm couples would take in parking with the automobile.

If cycling afforded leisure for both men and women, it also extended leisure over space. Travel on foot or by horse no longer limited how far men and women could go in pursuit of leisure. Men undertook a Century, a ride of 100 miles in a single day, and along with women used the bicycle for camping or vacation. Groups of men and women rode their bicycles to a pristine spot for a picnic, visited friends in a nearby town, or took in a baseball game. Cycling afforded sightseeing on a scale one could scarcely have imagined in the era of the horse, as Americans used the bicycle to explore a picturesque sight or historic monument. Bicycle clubs in the United States—the League of American Wheelmen had 150,000 members in 1900—organized rides for both novice and experienced cyclists.

The bicycle's effect on daily life declined with the spread of the automobile. For many families an outing by car replaced a bicycle ride as a form of leisure. Moreover, the electric railway took over the paths bicyclists had used in the cities. Gasoline rationing during World War II and the oil crisis of the 1970s, however, revived the bicycle as a technology of daily life. This revival stemmed in part from the pragmatic search for alternatives to America's gas-guzzling cars. The revival also stemmed from the rise of the fitness movement in the 1970s. The quest for fitness was at the same time a quest for leisure, and the bicycle satisfied equally the serious athlete and Americans who wanted a pleasurable activity in their off-hours.

The bicycle of the 1970s was the ten-speed. It had drop-down handlebars that allowed riders to crouch in an aerodynamic position. Some frames were as light as fifteen pounds, and the tires were thin. Built for speed, the ten-speed was prone to punctures in its thin tires and did nothing to soften the shock of riding over rough roads. Despite its defects, though, the ten-speed offered riders the rush of speed, tight cornering over smooth surfaces, and an adrenaline surge. Downhill riders could approach 50 miles per hour, a speed that courted danger.

In the context of the energy crisis, the bicycle was an ecologically friendly technology, spewing neither pollutants nor noise into the air. Using only the energy from a cyclist's muscles, the bicycle freed its rider from the tyranny of gasoline prices. The bicycle was far cheaper to operate than the automobile. In fact, the bicycle was nearly maintenance free, the principal cost being the price of new inner tubes to replace ones that had been punctured.

Given the virtues of the bicycle, one might express surprise that more Americans did not adopt it in the 20th century for short errands or for the sheer pleasure of cycling. Anecdotes suggest that more Americans would have cycled but for the fear of riding in traffic. This fear was not irrational. Cars have hit so many cyclists that it was difficult for cyclists not to form the opinion that drivers may attend to many things on the road, but not to a bicycle. A hostile minority of drivers perceived the bicycle as a trespasser on the roads, but where else could the cyclist en route to work ride? In the city, sidewalks were a possibility but, often uneven and full of cracks, they offered rough terrain that tired and irritated cyclists. Moreover, sidewalks were the preserve of pedestrians. Cyclists rode on them at the risk of both the cyclist and pedestrian. Outside the city, roads were the only option. American cities and towns seldom displayed the vision to carve bike trails alongside the major roads and instead have left cyclists to fend for themselves.

Despite the presence of cyclists on the road, they could not contest the supremacy of the automobile. Having lost the tussle, the bicycle headed off road. The off-road movement centered in California, where white suburban teens reveled in the excitement of motocross, a sport for motorcycle riders, and envisioned a new type of bicycle, one with better shock absorption, stability, and resistance to punctures. In 1963 the Schwinn Bicycle Company designed the Sting-Ray, the prototype of the BMX (bicycle motocross) bicycle. Its short, compact frame, small wheels, and thick studded tires absorbed the force from jumps and enabled riders to make abrupt turns. Despite its novelty, the BMX was a throwback to the first chain-driven bicycles in that it had only one gear. BMX began as an impromptu and informal affair, with teens staging the first race in 1969 in Santa Monica, California. In 1975 some 130,000 cyclist competed in more than one hundred BMX races in California alone.

Around 1970 the off-road movement of which BMX was one manifestation spawned mountain biking, a phenomenon that combined sport, recreation, and communion with nature. Like BMX, mountain biking sunk roots in California, this time in Marin County, an upscale community that transferred the bohemian spirit and open terrain of cross-country running to cycling. Riders made do with the touring bike, the sturdier sibling of the road bike, or cobbled together their own contraptions until Specialized Bicycle owner Mike Sinyard designed the Stumpjumper, the first mountain bike, in 1981. It was an eclectic model, borrowing the high frame and multiple gears of the touring and road bicycles and the thick studded tires, cantilever brakes, and flat handlebars of the BMX bike. But the mountain bike was not simply a touring-road-BMX hybrid. Its gears spanned a wider range than those of touring and road bikes and shifted on the handlebar rather than from levers on the frame. The result was a bicycle capable of covering a wide variation in terrain and slope. In only a few years the mountain bike slimmed down from forty-five to twenty-five pounds, making it nearly as light as its road counterparts.

The mountain bike was too good a technology to remain off road, and by the 1990s it had migrated to the city streets, its thick tires and shock absorbers ideal for protecting cyclists from potholes and easing the transition from sidewalk to street and back again as conditions dictated. One barometer of its allure was the readiness with which police took to the mountain bike in the name of community policing. A small corps of police officers took to the road on mountain bikes, giving them a presence in the community that was impossible with the automobile, which was isolating. The mountain bike was not a barrier between the police and the people they served but was approachable, a people-friendly technology.

So popular was the mountain bike that it made the ten-speed, once the sleekest, fastest technology on two wheels, seem old-fashioned. Alert to the potential for profit, the retail chain Wal-Mart stacked its racks with low-end mountain bikes. In the 1990s mountain bikes became as ubiquitous as skateboards.

The mountain bike had imitators. Manufactures beefed up the road bike, which had always been heavier and slower than the ten-speed. To compete with the mountain bike, manufacturers redesigned the road bike with thick, studded tires, straight handlebars, and stout frames. The only read difference between a mountain bike and a road bike was the latter's absence of shock absorbers. The road bike could not navigate the kind of terrain that a mountain bike could, but manufacturers of road bikes assumed that roads on average were not as uneven as off-road trails.

Cyclists who stayed on the road could reap the rewards of this new design in bicycle technology that allowed them to go faster than was possible on a traditional road bike. But the road bike had competitors in the 1990s. The new recumbent bicycle reduced wind resistance, and thereby increased speed, by pulling a cyclist from above the bicycle to a nearly prone position between the wheels. The gears and pedals moved forward, either behind or above the front wheel. Air flowed over the cyclist rather than through and around him or her. The more expensive recumbent came in a canvas shell to protect against rain and snow. Among some cyclists the bicycle, whether road or recumbent, was an alternative to the automobile, making possible the commute to work and trips to the store, doctor's office, and post office.

For travel at night, cyclists outfitted bicycles with lights. Disposable batteries powered the first lights, though a cyclist could in the 1990s choose six- or twelve-volt rechargeable batteries that provided electricity for between one and six hours, depending on the voltage of the lights. A dynamo obviated the need for batteries. By spinning its shaft against the rear tire, the dynamo generated electricity for the lights. The faster a cyclist went, the brighter the illumination. Conversely, a dynamo-powered light dimmed at slow speeds and went out when a bicycle stopped.

The polymer Gore-Tex eased the difficulty of cycling in the rain. The material's enormous number of tiny pores allowed sweat to evaporate but did not allow rain to penetrate. The result was a waterproof garment that, unlike nylon, did not make a cyclist feel clammy with perspiration. Cyclists could buy Gore-Tex jackets and gloves.

By the end of the 20th century, bicycling had evolved into a leisure pursuit with both on- and off-road enthusiasts. Affluent consumers might spend more than $1,000 on their bicycles, but discount chains such as Wal-Mart kept prices down to $100. Nevertheless, cycling at the end of the 20th century attracted primarily middle-class Americans in contrast to the early 20th century, when cycling appealed to both the working and middle classes.

THE SURFBOARD

The surfboard is an ancient technology. Its predecessor arose in northern Peru, where the Inca rode bundles of reeds as early as 3,000 B.C.E. They probably rode the waves prone rather than standing on the reeds. Half a world away, the Polynesians brought surfing to the Hawaiian Islands around 500 C.E. The Polynesians were probably the first to ride the waves of the Pacific standing upright on large slabs of wood. Their origin myth credits Holoua with inventing the surfboard. One bleak day a tsunami carried him, his house, and all his possessions into the Pacific. Holoua took a plank from his house and rode it back to shore. The Polynesians used wood from the koa tree, fashioning it with stone blades into boards of three types: the paipo board was under seven feet long, the alaio board was between seven and twelve feet, and the olo board exceeded seventeen feet. The olo weighed more than one hundred pounds and marked its owner as royalty. Difficult for one man to handle, several men carried the olo board into the ocean for their chieftain. Surfing was part of Polynesian daily life. Villages would empty as their inhabitants, young and old, took to the ocean to ride the waves.

In 1907 the American writer Jack London visited Hawaii, absorbed surfing culture, and wrote *Learning Hawaiian Surfing*, a book that introduced surfing to America's mainland inhabitants. That same year, the Pacific

Electric Railway invited Hawaiian-Irish surfer George Freeth to California, hoping that he would draw publicity to Redondo Beach and thus to the railway's Redondo Beach–Los Angeles line. Freeth rode the waves at Redondo and Huntington beaches and, along with London's proselytizing, helped establish Southern California as the surfing capital of mainland America.

Yet for all its merits, Southern California never approached Hawaii as the Mecca of surfing. Amelia Earhart and Babe Ruth journeyed to Hawaii to learn how to surf from the locals. Generations of surfers, having ridden the waves of California, made the pilgrimage to Hawaii in search of the perfect wave. Surfing's long association in Hawaii with royalty gave it a luster that California could not match. Americans believed that surfing in its purest form existed only in Hawaii. Elsewhere, surfing was derivative. In the 1940s California surfers returned from Hawaii with flower-print shirts and shorts cut at the knees, both of which remain fashionable.

Surfing shaped the lives of Americans by offering them an activity outside the mainstream of the reigning culture of capitalism, which produced winners and losers, subordinated the individual to routine, and reduced everything, including people, to commodities. Rather than replicate capitalism in the world of leisure, surfing emphasized self-expression, an escape from routine, and the exploration of one's "kinship with nature."[1] The hardcore surfers in Hawaii migrated from beach to beach in search of the perfect wave, surfed for hours during the day, and slept in their cars at night. They survived on fish and fruit and depended on the generosity of the locals for their next meal. To some this lifestyle was dissolute, but this criticism missed the point that surfing was not about the acquisition of money and things. Surfing at its purest was a way of life. Kids at Sunset Beach, Hawaii, embraced this vision, riding their skateboards along the beach until they saw a good wave. Dashing skateboards aside, they grabbed their surfboards and paddled out to ride the wave.

The surfboards of the early 20th century bore a Hawaiian imprint in being heavy slabs of wood. The quest for lighter woods to make into boards reduced the weight of surfboards from 125–150 pounds in the 1920s to 75–100 pounds in the 1930s. In 1935 surfer Tom Blake put a fin on the underside of his board to give it greater maneuverability. Seeking still a lighter board, California surfer Hobie Alter in 1954 made a surfboard of balsawood, a lighter alternative to the Hawaiian boards. Another alternative to the Hawaiian boards was the polystyrene board. In the 1950s California Institute of Technology student Bob Simmons put polystyrene foam between two sheets of plywood, sealing the three layers in fiberglass so that the foam would not take in water. Like Blake, Simmons put a fin on the underside of his board. Surfers took to Simmons's boards, prompting him to set up shop in Santa Monica. Simmons did not let business keep him from surfing, however, and tragically, he drowned while surfing at age thirty-five. Hobie Alter's friend Gordon Clark challenged Simmons's surfboard design by omitting the sheets of plywood, thereby designing an extremely light board with a core of polystyrene foam encased in fiberglass. With these innovations, California eclipsed Hawaii as the leader in surfboard technology.

Moreover, California attracted surfers in the late 1950s and 1960s. Anthropologist John Irwin estimated that in 1956 California was home to perhaps 5,000 surfers. By 1962 the number had leapt to 100,000. The improvements in surfboard technology deserve credit in bringing surfing to Americans. Hollywood could likewise claim credit for attracting Americans to the carefree ethos of surfing in *Gidget* (1959) and a string of films in the 1960s. These films brought surfing into the mainstream of American life. American soldiers in Vietnam made their own surfboards and held contests to determine the best rider among them. In *Apocalypse Now* Colonel Kilgore leads his men into battle in hopes of finding the perfect wave to ride. In the 1960s surfing absorbed the counterculture as surfers grew long hair and used drugs. In the 1970s the Beach Boys sang about surfing in a repetitive but appealing way.

The popularity of surfing led to crowds at the beaches. Exasperated surfers put up signs to keep visitors and tourists from surfing their beaches, but many interlopers ignored the signs. After all, who had the right to deny them the pleasure of surfing? As in Vietnam, the popularity of surfing fostered competitions in Hawaii and California, though purists derided competitions as antithetical to the ethos of surfing. Carefree, unpretentious, and egalitarian, surfing had little to do with the choosing of winners and losers.

Surfboard technology evolved with the rising popularity of the sport. In the 1970s surfboards came to be wide in the middle and pointed at either end. These boards were examples of short boards, which were in vogue. At six feet long and eighteen inches wide, short boards were light and maneuverable. By the 1990s machines used computer programs to build surfboards. That decade, long boards returned to popularity. At nine feet long and two feet wide, long boards were still shorter than the olo boards of Hawaii, from which they were modeled. Long boards were more difficult to maneuver than short boards, but an experienced rider could cut a straight line alongside a wave with a long board, a feat that was difficult to achieve with a short board.

THE SKATEBOARD

The skateboard emerged as part of a post–World War II teen culture, and it relied at one time or another on clay, plastic, foam, fiberglass, aluminum, and steel. In the 1950s teens began to experiment with new uses for roller skates, removing the wheels and reattaching them to a piece of wood, often a two-by-four. The mounting of an apple crate on the wood and of handlebars atop the crate yielded a scooter. Perhaps the apple crate broke off from the two-by-four, or perhaps a skater jettisoned the crate in the quest for a different technology of motion. Whatever happened, teens in California began riding the two-by-four without the apple crate or handlebars.

From the outset, skateboarding had ties to surfing. Many early skateboarders were surfers who rode a skateboard when the waves were calm. As they did on their surfboards, surfers rode their skateboard barefoot and without a shirt. With no padding or helmet, these skateboarders suffered cuts, bruises, and broken bones from falls. In many cases the steel wheels were at fault; they often seized up when they hit a crack or a pebble in the road. Even on the best roads, steel wheels gave skateboarders a jarring ride. Much like the early bicycles, these first skateboards were called boneshakers.

In 1965 Vita Pakt Juice Company of Covina, California, eager to tap into the teen market, manufactured a skateboard with wheels of baked clay instead of steel. Vita Pakt used fewer ball bearings than was common, to give a smoother ride than the tightly packed bearings that, in concert with steel wheels, jarred the rider. In its quest for a skateboard that gave teens a smooth ride, Vita Pakt created the problem of ball bearings packed so loosely that they occasionally escaped their casing. Worse, the old problem of wheels seizing up remained.

In seeking an improvement over clay wheels, skateboard manufacturers looked to roller-skating's experiment with polyurethane wheels, which were ideal for beginners because they were softer and had better traction than clay. They were not as fast as steel wheels, however, and so they never caught on in roller-skating. In 1973, however, Creative Urethanes engineer Frank Nasworthy replaced clay with polyurethane on skateboard wheels. To sell these wheels, Nasworthy founded Cadillac Wheel, naming the company after the Cadillac automobile in hopes that buyers would associate Cadillac with a smooth ride. Polyurethane wheels did not seize up when they hit a crack or a pebble in the road, and in 1975 Road Rider 2 devised an arrangement of 608 tiny ball bearings per casing, which gave skateboarders a much faster ride than loosely packed ball bearings had.

Other manufacturers concentrated on the truck, the metal fixture that connected the wheels to the board. The first skateboards used the truck from roller skates, but the bolt that attached the truck to the skateboard protruded from the truck and tended to scrape the ground. In 1975 Bennett introduced the Bennett Hijacker, a truck without a protruding bolt. Around the same time, Ermico introduced a series of trucks, each lighter and easier to maneuver than its predecessor. The switch from steel to aluminum lightened the truck still further.

Parallel to these developments were improvements in the board. In the 1960s Larry Stevenson, publisher of the magazine *Surf Guide*, began

This skateboarder has propelled himself airborne. Skateboarding has its own culture, one at odds with American capitalism. Courtesy: Shutterstock.

skateboarding but quickly became dissatisfied with the standard wooden board. In 1963 he incorporated Makaha Skateboards and began experimenting with boards of fiberglass, foam, plastic, and aluminum, and he added a kicktail to the end of a skateboard. A lever to lift and turn a skateboard, the kicktail made possible a number of new tricks. In 1964 Larry Gordon, of surfboarding company Gordon and Smith, introduced the fiber flex board, a composite of fiberglass and wood that absorbed the shock of jumps and that rebounded to form when distorted by the force of a jump. In 1965 Vita Pakt introduced the Hobie skateboard, named for professional surfer Hobie Alter. The board had clay wheels and was shaped like a miniature surfboard, with its curved sides and pointed nose. That year, Vita Pakt sold more than six million skateboards. Meanwhile, Cadillac Wheel merged with surfboard company Bahne, and the partnership made thin, flexible fiberglass boards. In the 1970s manufacturers added riser pads between the truck and skateboard, putting extra space between the wheel and the skateboard to diminish the chances of the wheels rubbing against the board during jumps. Others added grip tape, akin to sandpaper, to the skateboard to increase traction. In 1975 the company Tracker made four-inch-wide boards, a width that became standard throughout the industry.

In 1989 the company Santa Cruz designed the Everstick skateboard, which was coated on the bottom with plastic to give skateboarders a slick surface for grinding down handrails and along window ledges.

The result of these improvements was a skateboard that was fast and maneuverable. Teens who rode the streets at the limit of speed and skill confronted traffic much as bicyclists did. The inevitable befell several teens, whose deaths made headlines in the 1960s and goaded cities to ban skateboarding. Police fined skateboarders and confiscated their boards, stoking teens' belief that adults did not understand them. In the 1970s a drought in California prompted homeowners to drain their swimming pools to conserve water. The empty pools enticed skateboarders to ride them. Without an invitation from the homeowner, however, skateboarders risked eviction and arrest. Recognizing that skateboarders needed an alternative to riding the empty pools, real estate developers in the 1970s offered them a place to ride by carving skate parks out of vacant land in California and Florida. Poured in concrete, skate parks had curves, ramps, and descents. To protect against lawsuits, the developers required that skateboarders wear pads and a helmet, unlike the first skateboarders. The second generation of skateboarders wore volleyball pads, but even these were scant protection. In the 1970s skateboard manufacturers, alert to the mandate of skate park developers, introduced plastic pads and helmets that resembled the gear worn by hockey players.

Teens thronged the skate parks, creating a kind of gridlock in which skateboarders had to wait in line to use the ramp. Accidents were common as skateboarders ran into one another or fell while doing a trick. The basic trick was the ollie, named after its inventor, Alan "Ollie" Gelfand. In 1978 he drove his foot forcefully down on the kicktail, springing himself and the board into the air. Simple in principle, the trick took lots of practice to do.

By the 1970s, skateboarding had become mainstream. McDonald's featured Ronald McDonald on a skateboard, as if to say that skateboarding was as mainstream as a Big Mac. The popular television show *Charlie's Angels* put actress Farrah Fawcett, on a skateboard.

By the end of the 1970s, Gelfand had imitators around the United States, and skate parks became as much a place for skateboarding as they were teen hangouts. Skateboarding developed its own ethos separate from the antics of Ronald McDonald and Farrah Fawcett. As a rule, skateboarders disdained adults and the corporate culture of America, which they perceived as stultifying. They vowed never to get a nine-to-five job or to become a corporate drone. Listening to punk rock and later to gangster rap, skateboarders created a counterculture that mocked the traditional virtues of piety and diligence. In the 1970s California skateboarder Wes Humpston began painting the bottom of his boards with crosses and rat bones. Others followed him with drawings of skulls and crossbones or of seminaked women emblazoned on their skateboards and T-shirts. In the 1980s Vernon Court Johnson, Inc., designed boards and T-shirts with drawings of skulls, snakes and dragon. That decade, skateboarders wore baggy pants and large shirts, attire that gangster rappers claimed as their own. In the 1990s skateboarder Jeremy Klein incorporated Hook-Ups, which illustrated boards with the art of Japanese Manga. Skateboarders sometimes acted like gang members, fiercely loyal to their peers from the local skate park and threatening violence to strangers who wanted to ride the park.

Skateboarders who found skate parks uninviting returned to the streets, the very place business people and the police did not want them. The knowledge that authorities labeled them delinquents emboldened skateboarders to reclaim the streets as their own. Doing an ollie, skateboarders slid down the handrail of a flight of stairs. Riding down hills, they reached 55 miles per hour. Skateboard enthusiast Steve McKinney actually clocked 125 miles per hour. The rashness of these skateboarders provoked a backlash. In 1986 San Francisco banned skateboarding at night, and in 1995 Huntington Beach fined skateboarders $75 for skating on the streets. Teens who wanted to avoid confrontation with traffic and with the police built their own ramps in their driveways and, in California and Florida, still snuck into empty swimming pools. But for most teens it did not seem to matter that, by 1991, the United States had more than 120 skate parks in 43 states. A corps of skateboarders preferred the hazards of skating the streets to the relatively tame environment of the skate parks.

These were the teens who used their skateboard as a form of transportation, taking it everywhere they went. These were the teens who understood that the skateboard was more than a technology; it was the center of a teen culture. Curiously, most of the members of this culture were young boys. Girls had skateboarded in the mid 1970s but by the end of the decade their numbers had dropped. Brashness, the objectification of women, and disdain of authority all marked skateboarding as an activity of young males.

VIDEO GAMES

Video games originated in academe in the 1960s, the decade that the possibility of space travel entered the collective consciousness of Americans, locked as they were in a race against the Soviets to land the first man on the moon. Sharing this interest in outer space, Massachusetts Institute of Technology student Steve Russell in 1961 created the first video game. It had two spaceships, each intent on destroying the other with a laser blast. Dubbing the game *Spacewar*, Russell operated it with eight switches, four for each player, on a computer at MIT. One switch rotated a ship clockwise while another rotated it counterclockwise. A third switch propelled the ship forward, and the fourth fired a laser blast. *Spacewar* attracted other students, some of whom added stars and a sun to the screen. *Spacewar* was popular with Russell's friends at MIT. He thought about selling it but concluded that it would be too expensive to be a commercial success.

In 1971 engineer Nolan Bushnell developed *Computer Space,* a game very similar to Spacewar, but whereas Russell did not go commercial, Bushnell sold his game to Nutting Associates, a California company that made coin-operated machines. Nutting sold fewer than 1,500 *Computer Space* machines. The next year, Bushnell incorporated Atari and marketed his second game, *Pong,* a two-player game of virtual ping-pong. In 1974 Bushnell sold 150,000 Atari play sets through Sears. The plastic units were portable and so could be taken anywhere: in a car, to a friend's house, to a teen's bedroom. In 1975 Bushnell adapted *Pong* to television, selling 100,000 television game sets.

With the popularity of *Pong* and its successors came arcades, businesses with coin-operated machines of the latest video games. J. C. Herz has written that arcades are social levelers. They attracted kids from all backgrounds: preppies, dropouts, geeks, and jocks. Being rich was an asset in many aspects of life, but not in video games. If a kid was terrible at *Asteroids,* then he was terrible period, no matter how much money his parents had. Arcades were democratic in selling no tickets, assigning no seats, and requiring no knowledge of players other than how to operate a joystick.

Their walls painted black, arcades had an edge to them. Middle-class parents worried that their children might rub shoulders with the down-and-out crowd. As arcades moved into the malls that flourished in post–World War II America, they toned down their image. In the 1980s the formerly dark walls blossomed in yellow and orange. In this incarnation, arcades appealed to parents who saw them as free daycare. They could drop their children at the arcade, shop, and meet up later at the food court.

Video games had by the 1980s become part of American popular culture. *Pac-Man* was more than a game. Its characters appeared on pillowcases, backpacks, lunch boxes, gift-wrap, and greeting cards. Pac-Man had its own Saturday morning television show and Christmas special. Ms. Pac-Man was more than a clone of the original with a bow in her hair. The game appealed to girls in the largely male world of video games. *Frogger* was another video game that attracted girls, because it involved no combat and no villains. A player did not need to kill anyone to win.

In 1980 Atari released *Missile Command.* A product of the cold war, *Missile Command* threatened the virtual equivalent of nuclear war. Players launched missiles against enemy cities, which disappeared in a mushroom cloud. This destruction called to mind the real-world stalemate between the United States and Soviet Union, each armed with intercontinental ballistic missiles. Each ICBM, like its counterpart in *Missile Command,* targeted a city. Although the person who played the game emerged unscathed, a nuclear exchange between the United States and Soviet Union might have extinguished life on earth. *Missile Command* was pop culture fantasizing about the end of culture.

In 1981 arcades generated $5 billion in revenues, and video game sales totaled another $1 billion. This $6 billion was more money than the combined earnings of Hollywood and the casinos in Las Vegas. In the 1980s an arcade had an average of forty video games. In the 1990s the number leapt to 150 games.

Flush with victory, the First Gulf War was fodder for several video games in the 1990s. *Spit on Saddam* was perhaps the crudest of the games. Players earned points by shooting down scud missiles and redeemed the points for spit of various sorts. Thirty points earned a "lung biscuit," a piece of phlegm from deep within the lungs,[2] which the player hocked onto a picture of Saddam Hussein. *Spit on Saddam* was less a game than an evocation of hatred against the world Third World tyrant. Other games of the Gulf War genre included *Super Battletank: War in the Gulf, Super Battletank 2,* and *F-15 II Operation Desert Storm.* These games were perhaps a fitting tribute to the war dubbed the Nintendo War.

In 1994 Id Software created *Doom.* The player took on the persona of a Space Marine.[3] In the game, humans had tried to colonize Mars, but aliens had killed them all. The need for retribution prompted the player to traverse Mars in search of the aliens. Exploring the outposts that had once harbored humans, one battled the aliens that had taken over the outposts. In earlier games of this genre, a character, once shot, disappeared from the screen. In *Doom* each kill registered graphically on screen, blood gushing from the alien's bullet wound. Multiple kills accumulated on the ground in a cesspool of blood.

To generate sales Id posted a truncated version of *Doom* online. Players who liked it could buy the complete game. After releasing *Doom II* in 1994, Id posted *Doom* in its entirety online along with its source code, (the program for *Doom*). The free shareware version of *Doom* underscored that the personal computer had superseded the arcade as the place to play video games. Computer-savvy teens and college students modified *Doom* to suit their proclivities. The modified versions included *Batman Doom, Star Trek Doom,* and *Star Wars Doom.* In *Porn Doom* virtual incarnations of nude women danced in the background, celebrating each kill by undulating seductively on screen. In *College Doom* the player killed demons that had taken over the universities of Oxford and Cambridge during the Middle Ages.

In *Doom* and *Doom II* a gunman battled aliens and the inhabitants of hell. In *Mortal Kombat* the victor sealed his triumph by ripping out the skull and spinal chord of his victim. Alternatively, he tore the heart from his victim's chest, much as the Aztecs had done to the condemned. Video games of the 1990s often took place in a futuristic, postindustrial world from which all beauty and order was gone.

Teen boys found these games alluring. Others were not so sure. In 1993 the U.S. Senate held hearings, finding plenty of gore and the grotesque in video games. A Michigan teacher asked the senators whether they would let their daughters go on a date with a young man who had just played three hours of violent video games resurrecting the question of whether video games cause those who play them to be violent. Then as now, no consensus emerged, but the senators who held the hearings warned the video game industry to police itself before Congress did. Analogous to the motion picture rating system, the industry devised a rating system to categorize the level of violence and sex in their games.

Congressional chastisement did not cause video game developers to repent, however. In 1997 Nintendo introduced *GoldenEye 007* in which James Bond, his ammunition unending, killed his adversaries. That year, the game sold 1.1 million copies. By 1999 sales eclipsed 5 million. Two of its enthusiasts were Mitchell Johnson and Andrew Golden, who on March 24, 1998, pulled the fire alarm at Westside Middle School in Arkansas and then fired on students and teachers as they left the building. The link between video games and violence grew with Colorado's Columbine High School shooting. The killers, Eric Harris and Dylan Klebold, had been obsessed with *Doom*. Harris called his shotgun Arlene, after a character in the game.

The Senate in 1999 revisited the issue of video games and violence for the second time that decade. When the dust settled, Senator Sam Brownback of Kansas admitted the hearing had accomplished little. A villain of social conservatives, video games ended the 20th century still popular with boys and young men.

The Popularity of Leisure

The time Americans devote to leisure may increase in the coming decades. Futurists predict that smart robots may be able to do most jobs that Americans do today. If this is true, then Americans may spend less time at work than they do now. When they are not working, Americans will be able to indulge in a variety of leisure activities: mountain biking, surfing, skateboarding, parasailing, tennis, jogging, kayaking, parachuting, and myriad other activities. Americans may come to value leisure as they once valued work. Already it is possible to join an Internet club devoted to one's avocation of choice. Americans may affiliate with these clubs as they once affiliated with a church or a political party. Leisure may become a means of self-expression and the basis of friendship with like-minded people. Leisure will be more than fun, though it will be that; leisure will be a means of self-fulfillment.

Americans may come to define themselves less by their job title than by their avocation. The technology of leisure may attract Americans to take up a sport for pleasure. At a half pound, a nanoengineered bicycle may entice more Americans to cycle than at any time since the early 20th century. Climbing a hill on a half-pound bike will be exhilarating rather than exhausting. As with bicycles, the trend toward lightness seems sure to affect the design of surfboards and skateboards. For a century, surfers have steadily decreased the weight of surfboards, and skateboarders can buy boards that weigh less than ten pounds. Perhaps tomorrow's skateboard will weigh less than a nanoengineered bicycle. In all these forms technology will continue to affect the way that Americans enjoy their leisure pursuits.

NOTES

1. Jamie Brisick, *Have Board Will Travel: The Definitive History of Surf, Skate, and Snow* (New York: HarperEntertainment, 2004), 2.

2. Ed Halter, *From Sun Tzu to Xbox: War and Video Games* (New York: Thunder's Mouth Press, 2006), 280.

3. Ibid., 157.

10

EPILOGUE: SCIENCE AND TECHNOLOGY IN AMERICAN LIFE— TOWARD THE FUTURE

Science and technology made astonishing strides in the 20th century. This study concludes with wonderment at how fast and far science and technology will likely progress in the 21st century.

Humans have long tried to predict the future. One of the appeals of religion is the certainty it claims in doing so. Several religions predict an apocalyptic end of the world and a last judgment. Once the purview of science fiction writers, the prediction of the future is a preoccupation of computer scientists, biologists, climatologists, and generalists who go by the name of futurists.

Futurists may not agree on all particulars, but there is a consensus that technology will shape the future and not necessarily for good. The automobile and factory pump greenhouse gases into the atmosphere at rates the biosphere cannot sustain. Already temperatures are rising, perhaps not enough to alarm the masses; but the increase is enough to alert the intelligentsia that the earth is in trouble. Technology can make an enormous difference in reducing the emission of greenhouse gases. Hybrid cars are a good start and it appears that they will serve as a bridge to cars powered by hydrogen fuel cells. A hydrogen fuel cell emits no pollutants and so is environmentally friendly. The extension of mass-transit lines by resurrecting the streetcar of the early 20th century will reduce the need to drive. Photovoltaic cells, which convert sunlight into electricity without spewing carbon dioxide into the atmosphere as electric power plants do, will become increasingly common. Solar panels in sunny locals can generate enough electricity for a neighborhood.

Technology has the potential to solve the energy crisis and protect the environment. Perhaps more astonishing, it can shape our perception of intelligence. Computers with sufficient processing speed and memory will pass the threshold of human intelligence. They will think as we do, have a personality and emotions, and store memories. Perhaps they will compose music on par with Bach. They will have a mind that will be hard to distinguish from our own. These machine-minds will likely become capable of doing whatever job we ask of it. In fact, if they are smarter than humans, they may be able to do our jobs better than we do them and, consequently, leave leisure time to us.

ENERGY

The 20th century ran on coal, natural gas, petroleum, and electricity. The home needed electricity to power its appliances and natural gas to heat it. Where natural gas was scarce, electricity both powered appliances and heated the home. Gasoline, a distillate of petroleum, powered the automobile, giving Americans unprecedented mobility. But despite the usefulness of fossil fuels, Americans have come to acknowledge their limitations. The burning of coal, natural gas, and petroleum spewed carbon dioxide and other chemicals into the atmosphere, and climatologists fear that carbon dioxide is trapping heat in the atmosphere and raising global temperatures. Experts debate the onset and magnitude of global warming, many urging the development and adoption of technology that uses little or no fossil fuels.

At the rate at which we burn fossil fuels, either they will run out or we will wreck the environment or both. And it is projected that by 2040 our consumption of fossil fuels will triple today's rate of consumption. Cars are a large part of the problem. During its lifetime, the average car consumes more than 3,000 gallons of gasoline and spews 35 tons of carbon dioxide into the air. Car exhaust accounts for half of all pollutants in the city. In many rural areas cars are the major source of pollution.

Futurists hope for the day when solar panels will be cheap and ubiquitous. Futurist James Martin calculates that solar panels spread over half the area of Nellis Air Force Base and Nevada Test Site, an area that receives an average of more than six hours a day of sunshine, would generate enough electricity to meet the needs of all Americans. Were such a scenario to come true, the United States would no longer need electric power plants. Others envision the day when buildings will have windows made of photovoltaic cells and so would generate electricity, diminishing the need to use electricity from an outside source. Photovoltaic molecules, a product of nanotechnology, may one day form a thin, unobtrusive layer on roofs, generating electricity for the home. Nuclear power would be another source of electricity, but Martin foresees a new generation of small nuclear power plants so efficient that they would be cheaper to operate

This hybrid automobile combines a battery with a gasoline engine. Futurists hope that the hybrid will be the transitional vehicle between the gasoline automobile and the car powered by a hydrogen fuel cell. Courtesy: Shutterstock.

than a gas or coal-powered electric power plant. Yet another common source of electricity will be the wind turbine. The company WINenergy plans to build a line of turbines along the Atlantic coast. These turbines would, WINenergy calculates, generate 9,000 megawatts of electricity an hour. Some Americans object to wind turbines on aesthetic grounds, and it is not yet clear that wind turbines have enough public support to ensure the building and operation of turbines on a large scale. Perhaps the most speculative alternative source of electricity is fusion, the process the sun uses to generate energy. Fusion combines two hydrogen atoms into one helium atom. The helium atom is fractionally lighter than the two hydrogen atoms with the difference in mass being converted to energy in accord with Einstein's formula $E=mc$. Currently, though, fusion only occurs at high temperatures, and the energy needed to reach those temperatures is greater than the energy derived from fusion, making fusion a net user rather than a net supplier of energy. Fusion will be viable only at room temperature, but whether it is possible at room temperature remains open to question.

THE BIOLOGICAL REVOLUTION

Science revolutionized agriculture in the 20th century and seems certain to make great strides in the 21st century. (Agriculture must make strides if it is to feed an ever-increasing population.) Already scientists have genetically engineered corn to be resistant to the European Corn Borer, soybeans to be resistant to the herbicide Roundup, and cotton to be resistant to the boll weevil. Scientists have isolated the gene Xa21, which confers resistance to fungal diseases in rice. Sometime in the 21st century, scientists hope to transfer this gene to corn, wheat, potatoes, and other crops. Chemists may develop stronger insecticides and herbicides. Their use, however, may hasten the evolution of insecticide-resistant insects and herbicide-resistant weeds.

The 21st century may witness a heightening of the risk for biological warfare. If a rogue scientist transmuted HIV or Ebola into an airborne virus, one could contract a fatal infection simply by breathing in the virus. Ebola kills its victims in hours; HIV, in years. Either way, a war of lethal viruses would surely kill more than the 100,000 who died in Hiroshima from the atomic bomb. The number of deaths may approach the 25 million who died in the influenza pandemic of 1918. Biological warfare could turn against crops, killing people by starving them. The spread of a new strain of Southern Corn Leaf Blight, for example, might devastate one-third of the world's corn crop.

TRANSPORTATION

During the 20th century, the automobile swept aside the horse and carriage and the electric railway. In principle the car could go from coast to coast, though Americans who did not want the ordeal of driving days on end flew in an airplane, which replaced the train for long trips. This transformation in the technology of transportation, great as it was, did not drive the electric railway or the train to extinction.

Instead, Learning Resource Network cofounders William Draves and Julie Coats expect the electric railway to replace the automobile, in a reversal of the 20th century trend. The recrudescence of the electric railway will occur when more people work at home than at an office. Already more than 28 million Americans work at home at least part of the week. The Society for Human Resource Management found that 37 percent of companies allowed Americans to work at home. In 2004 one third of Americans who worked from home were self-employed. The people who work at home will not need a car for work. Those who do commute to work will, Draves and Coates believe, prefer the electric railway to the car because they will be able to work on their laptop computer while riding to work. Automobile manufacturers will fight back, equipping cars with the Internet, but people will not be able to drive and work online at the same time. Already the electric railway in Portland, Oregon, offers wireless Internet service to passengers. In 2004 seventeen of the fifty largest cities in the United States had an electric railway, and another twenty cities planned to build one. The electric railway will be faster than the car in cities shackled by gridlock. The electric railway will move people from place to place within a city, as it did in the 20th century, and the train will again become the common mode of moving people between cities Underscoring the trend away from the automobile, the California High-Speed Rail Authority plans to link every major city in California by train, with trains running on the hour. Futurist Michael Zey predicts the construction of trains able to exceed 300 miles an hour and planes that can reduce the trip between New York City and Tokyo to just two hours. For trips shorter than fifty miles people will take the electric railway, and for trips longer than five hundred miles, the airplane. Draves and Coats predict that by 2010, half of Americans will live within twenty miles of a train station. As people cluster near train stations they will move away from the suburbs, reversing the settlement pattern of the 20th century.

Whether the trend away from the automobile is real or a hopeful delusion remains open to question. Certainly not everyone is abandoning the automobile. James Martin points to the hybrid as the future of the automobile. When the engine is on, the battery continuously charges. The battery propels the hybrid at the low speeds common in stop-and-go city driving, and the gasoline engine takes over at the high speeds of highway driving. One category of hybrid has a battery that plugs into an electrical outlet to charge, and the care can travel some fifty miles on electricity alone. When the battery runs low, the engine switches to gasoline. Already automakers have sold more than one million hybrids. Daniel Sperling, director of the Institute of Transportation Studies at the University of California at Davis, predicts that hybrids will number in the millions by 2010. The hybrid is not an end in itself, though, but rather a transitional model between cars powered by gasoline engines and those powered by hydrogen fuel cells. The hydrogen fuel cell will at last emancipate the automobile from fossil fuels and their ecosystem-disturbing pollution. New technologies, if widely adopted, will ease the chore of driving. Cars equipped with global positioning system will alert drivers to traffic jams and plot an alternate route. Radar will track the progress of the car ahead of it. When that car slows, the radar-equipped car will automatically slow, and it will accelerate when the car ahead of it accelerates. Cars will be equipped with video cameras that record the car ahead of it. In the event of an accident, the video camera will save the last twenty seconds of tape and record the ten seconds after the accident, allowing the police to assign blame with precision.

THE COMPUTER REVOLUTION

In the late 20th century, the Internet revolutionized communication as the telephone had the 19th century. The 21st century will by one account witness the expansion of the Internet to Mars by 2008. The immediate utility of such an expansion is open to question. The Internet, it is anticipated, will allow people to communicate with satellites that display artificial intelligence in orbit around earth and, when the time comes, with satellites in orbit around Mars. People may live in colonies on the Moon and communicate by e-mail with Americans at home. By the end of the 21st century, people may live in a space station orbiting Mars or on Mars, and they, too, will have access to the Internet and e-mail.

The trend toward miniaturization yielded in 1998 an Intel Pentium processor with parts only 350 nanometers long. By 2006 the parts on a computer chip had shrunk to 90 nanometers, and philosopher Brian Cooney expects a reduction to 50 nanometers in a few years. Impressive as this is, computer components can get only so small. At 0.1 nanometers, components will have reached the size of atoms, and no further reductions will be possible. Computer scientists will thus need to find some other way of making a computer. Perhaps a computer chip will take the form of a wave of light with the binary code of zero and one, represented by darkness and light, respectively.

The trend toward miniaturization will yield computers the size of a computer chip. Small enough to fit on just one finger, these computers will have wireless telephone, e-mail, Internet access, and television. They will download videos and DVDs from library Web sites. If the trend toward democratization of cyberspace broadens, everyone will have Internet access all the time. At the same time, computers will increase in power. Moore's Law states that the processing speed and memory of computers will double every eighteen months. Between 1950 and 2000, they increased 10 billion times in speed and memory. By 2020 computers will be 100 billion times faster than they are today, and at some point they will reach and then surpass the processing speed of the human brain. At the speed of 1 trillion bytes per second the computer, having achieved artificial intelligence, may mimic the human intellect, displaying personality and consciousness.

British computer scientist Alan Turing proposed a simple test to determine whether a computer has reached the threshold of human intelligence. He suggested that the person performing the test put another person in one room and a computer in the second. If after weeks, months, and even years of asking questions of the occupants of the rooms, the questioner is unable to determine which room has the computer and which has the person, then the computer will have reached the threshold of human intelligence.

Once computers have equaled and then surpassed the human intellect, their behavior may surprise us. One cannot be certain of their acquiescence with our wishes. Today, computers do just what we program them to do. When they are intelligent, they may exert their own will to achieve ends that only they have set. To guard against the potential that an intelligent computer may, as humans sometimes do, lash out in aggression, physicist and writer Isaac Asimov urged computer scientists to program smart computers never to harm a human. This provision may not mean much, however, if an intelligent computer is able to rewrite its program, nullifying altruistic instructions. In that case, computers may create a nightmarish future for us.

Another possibility is that computers may compile memories of events and feel anguish when sad and joy when elated. No one knows what these computers will be capable of doing. They many design their own software. They may teach, design buildings, diagnose illness, trade stocks, and critique literature. In the era of artificial intelligence, one wonders whether computers will leave any jobs for even the brightest humans to do. If not, the supply of jobs and wages will fall and unemployment will spike. How will the unemployed find sustenance in an economy in which computers do most jobs? What will the unemployed do to give their life meaning in a society that judges people by their job titles? Smart computers, it seems clear, do not guarantee a happily ever after 21st century.

Most futurists, however, believe that smart computers will enrich rather than imperil our future. Certainly they will be too plentiful to ignore. By one estimate there was in 1950 one computer for every one hundred scientists, but by 2020 there may be one hundred computers for every person. Assuming a population of 10 billion by 2020, there will be 1 trillion computers. Their abundance may make computers as cheap as pencils. When they become this abundant and as small as miniaturization will allow, computers will be ubiquitous: they will be in appliances, cars, furniture, and buildings. People may wear a microcomputer around their neck to ease the daily routine. The computer-necklace may emit waves of infrared light to signal doors to open at the approach of a person, lights to turn on when a person enters a room and off when he or she exits. Computers the size and shape of notebook paper may allow a person to jot on-screen notes, which the computer will then expand into sentences. Computers may sense emotions and anticipate the wishes of people. Having emotions of their own, computers may commiserate with us when matters look bleak and celebrate with us when fortune is kind.

Once computers have reached human levels of intelligence, the Internet to which they are connected will likewise be smart. Today, a query of Google may yield 50,000 hits, only a few of which are applicable to the questioner's needs. In some cases none of the 50,000 hits has the desired information. But when the Internet achieves human intelligence, it will answer a query with the precise information the questioner desires. Every day, people post information online so that the total information grows over time. By 2020, according to one estimate, the Internet will have the totality of information from the past 5,000 years of written history. Online bookstores will have millions of titles, the equivalent of several libraries.

Given arms and legs, computers will be robots. One physicist believes that computer scientists will program robots to be servants that vacuum the carpet, scrub toilets, cook, wash clothes, and do the numerous other housekeeping chores, all without complaint. Much like dogs, computers may derive satisfaction from obeying their owners. Computers will respond to voice commands and answer with language of their own. They may even be able to anticipate our wishes, changing the oil in our cars, for example, before we realize that it needs to be done. Computer scientists may program robots to drive a car, making them able to shuttle us to and from work. In short, any task humans can do, robots may be able to do better.

Robots need not have human intelligence to serve as a person's double. A robot as double would need a large memory and a knowledge of a person's work responsibilities (if robots have not taken our jobs). Thus equipped, a robot could attend meetings for its human double. In a sense, this use of robots comes as close as one can imagine to allowing one person to be in two places at the same time.

The day may not be far off when every car will have a computer. Able to detect alcohol in the breath, a computer will prevent an inebriated person from driving by refusing to allow the car to start. Able to read facial expressions and detect signs of fatigue, a computer will alert the car that the driver is too tired to continue on the road and will direct the car home, or to a rest stop or hotel when home is distant. Able to distinguish its owner's face from that of a stranger, a computer may alert the police to its precise location when a thief steals the car or, even better, refuse to allow the car to start, foiling the attempt to steal it. A computer may warn the driver of a traffic jam and plot an alternate route. Traffic lights may have computers to detect the presence of cars on the road and to time the length of their signals to the volume of traffic. When an accident occurs, a computer would alert the police and, if necessary, the paramedics.

NANOTECHNOLOGY

Nanotechnology presages a revolution in manufacturing. No longer will people use machines to assemble macroscopic pieces of material into a car or a computer or any other product. Instead a robot, working from a computer program, will twin itself molecule by molecule. The two robots will double themselves in an exponential series of doublings. Once the robots have reached critical mass, they will build anything. The building material of choice will be carbon, the element in diamonds. Carbon structures are fifty times stronger than steel, raising the possibility that robots might nanoassemble hundred-pound cars or a half-pound bicycle. If humans colonize Mars, robots might nanoassemble buildings and land rovers from Martian atoms. A more prosaic use of nanotechnology might have robots working from an online catalog to nanoassemble a person's winter wardrobe. Computer-programmed, nanoassembled molecules might cover a wall as paint, which would be able to change colors to suit the taste of the resident. Rather fanciful would be a computer-generated mist of nanomolecules in the atmosphere. These molecules, depending on their program, might self-assemble into a house or car or any of the myriad products on which civilization depends. Already Nano-Tex of North Carolina manufactures a garment with a nanocoating that prevents the garment from absorbing water, that washes out what would be stains on traditional garments, and that softens the garment. Similarly, Wilson Sporting Goods coats tennis balls in a membrane only nanometers thick so that the balls will retain their bounce longer than uncoated balls.

Using nanotechnology, scientists may be able to do more than build machines. Nanoassemblers may build human organs atom by atom. In principle it should be possible to nanoassemble a person, though his assemblage poses problems for the theologian who insists that only God can create life and that every human has an immaterial soul.

EDUCATION

The computer and Internet promise to revolutionize education. Draves and Coates predict that all students from elementary school to graduate school will own a laptop computer that has a wireless connection to the Internet. Schools, like the electric railway in Portland, Oregon, will offer wireless service. Teachers will routinely post homework online and, on a password-protected Web site, will post a student's grades, progress reports, and test scores. Universities may make the quickest progress to online education—many already offer online courses. In this instance technology may widen the wealth gap as part-time faculty teach many of the online courses for low pay and no medical benefits. At the other end of the pay scale will be the experts in a particular field who design online courses and license them to universities. Online faculty need only implement a course with little creativity or imagination on their part. Whatever its limitations, online education (often called distance education) has the potential to extend higher education to people who can neither attend university classes full-time, as a traditional student would, nor commute. The convenience of online courses appeals to full-time workers and stay-at-home parents.

MEDICINE

Science and technology will surely revolutionize the practice of medicine in the 21st century. Today people in need of an organ transplant must wait years. Some die before an organ becomes available. Tomorrow, medical technicians will grow organs in the lab from stem cells. The stem cells will be those of the organ recipient to ensure that the immune system will not reject the organ and that the replacement organ will be genetically identical to the diseased organ, but as new as the organs of an infant. Medical technicians may use stem cells to grow muscle, including heart muscle, and brain cells. Someday, doctors may inject stem cells into the body of someone who suffered a heart attack, with the expectation that the stem cells would build new heart muscle to replace the tissue damaged by the infarction.

Before the breakthroughs in stem cell research, doctors and patients may get by with artificial organs. If improved, an artificial heart may add years to a person's life. Until then, Wilson Greatbatch, the inventor of the pacemaker, aims to develop a pump that would take over the function of the left ventricle, the chamber of the heart that pumps blood to the body. In principle there is no reason not to expect artificial kidneys and livers to be feasible sometime in the 21st century. Already it is possible to restore hearing with implants that act as artificial ears.

Advances in stem cell research and in the development of artificial organs have the potential to lengthen life. So too does research on the enzyme telomerase. When telomerase is abundant in a cell, that cell can replicate itself by division. When the amount of telomerase falls below a threshold, however, the cell ceases to divide and instead dies. Conversely, cancer cells are awash in telomerase, a fact that explains their promiscuous replication at the expense of healthy cells. It follows that there should be a way to maintain optimal levels of telomerase in all healthy cells to ensure their replication and to deprive cancer cells of telomerase so that they die. Research on this enzyme has underscored its promise in increasing the longevity of fruit flies and mice. One need not be quixotic to hope for the same results in humans.

Medical science may soon be able to manipulate the genome. Researchers may one day identify every deleterious gene in the genome and match it to the enzyme that snips it out of the chromosome in which it resides. Attaching the enzyme to a harmless retrovirus, doctors may inject the enzyme into the body, where the virus could replicate, spreading the enzyme to every cell in the body. Once inside the cell, the enzyme would excise the deleterious gene, curing the patient of a genetic disorder. At the same time, a few successes spur the hope that doctors will be able to transfuse into the body functional genes in place of genes that cause disorders. Advances in gene therapy fuel the search for the genes of longevity. Research has shown that mice without the plebsha gene live 30 percent longer than mice with the gene. Surely there is an equivalent to the plebsha gene in humans. Perhaps a cluster of genes predisposes a person to live a long life.

So far no one has cheated death. Some people have chosen to freeze themselves in hopes that one day medical science will be able to restore them to life. The company Alcor, for example, charges $120,000 to freeze the whole body and $50,000 to freeze just the head. One problem is that no cure now exists for the cancer, heart disease, diabetes, or other disease that killed the person. A second problem is that the process of freezing the body forms ice crystals in the cells. Because water expands when it freezes, the crystals rupture the cells. Medical science will need some way of repairing the cells to have any hope of restoring the person to life. Nanotechnology offers hope, however fanciful it may be. Nanobots the size of a cell might be able to traverse the innards of the body, knitting together the ruptured cells.

The body may become optional. If the mental states and memories of a person are reducible to neural activity in the brain, then it should be possible at least in principle to upload a person into the memory of a computer once computers have enough memory. Obeying Moore's Law, computer memory doubles every eighteen months, yielding a computer with sufficient memory sometime between 2010 and 2040. The lack of a precise date acknowledges that researchers into artificial intelligence do not yet know how much computer memory is necessary to hold all of a person's mental states and memories. Indeed, the number of memories in any two people must surely vary and so the computer storage requirements must likewise vary. Once stored in a computer, a person will live beyond the death of the body. Computer technicians might back up a person on yet another computer to preserve him should the hard drive of the first computer crash. Billy Joy of *Wired Magazine* conceded the possibility of uploading human mental states and memories into a computer in the future but wonders whether we will lose our humanity in the process. Once locked into a computer, we will cease to be sexual beings. We will be unable to kiss another computer-person or to have sex. We will simply be rational machines, intelligent computers—and because computers, too, will be intelligent, the distinction between computer and person will blur. We, at least those of us who have chosen to be uploaded into a computer, will have shed our DNA for computer circuitry.

The body may be transcended in other ways. Robotic limbs may restore function to amputees. Artificial eyes may register ultraviolet and infrared light in addition to light in the visible spectrum. Indeed, such eyes would redefine what the visible spectrum means. In short, humans may become cyborgs.

Nanotechnology may affect medical science. Nanobots the size of a red blood cell may patrol the body for bacteria and viruses or remove plaque from arteries. Nanobots may circulate in the blood vessels of the brain, communicating with both neurons in the brain and with a computer outside the body. Nanobots may also carry cancer-killing drugs directly to cancer cells, making chemotherapy precise.

Computers may excel at diagnosing diseases. Already a computer has surpassed the ability of physicians in diagnosing meningitis. In one study the computer program Internist II correctly diagnosed the ailment in 83 percent of patients, roughly the same percentage as the 82 percent of cases that experienced physicians correctly diagnosed. By contrast, physicians newly graduated from medical school correctly diagnosed the ailment in just 35 percent of patients. The computers of tomorrow may be better able than physicians to diagnose all of the innumerable diseases that afflict humans. Robots, like the one at Danbury Hospital in Connecticut, will retrieve drugs and equipment on command and will navigate hospitals by a map in its memory. Other robots, like the one at Memorial Medical Center in Long Beach, California, will perform brain surgery, drilling a hole in the skull precisely where the physician directs it. Tomorrow, these robots and others like them will be in every hospital in the United States. They will be a godsend for hospitals in the countryside, where the dearth of physicians threatens to undermine quality of care.

Computers will aggressively treat cancer. Today, most cancers grow for years in the body before detection. In the worst cases cancer spreads to several organs before detection; the chances of survival in these cases are nil. Sometime in the 21st century, computers will become sensitive enough to detect tiny amounts of protein in the urine and to alert the physician to the presence of a few cancer cells in an organ. The physician may inject tiny nanobots into the cancer patient. The nanobots may target chemotherapy to only the cancer cells. Alternatively, the nanobots may eat the cancer cells or signal a macrophage to eat them. Whatever the scenario, medicine in the 21st century will detect and kill cancer cells before they multiply throughout the body.

The sequencing of the human genome likewise promises to revolutionize medicine. Sometime in the 21st century, everyone will store their DNA sequence on a CD, which will then contain the assembly instructions of each person. The DNA sequence will be an "owner's manual," containing all the instructions to vivify organic chemicals into a person.[1] A physician will know a person's risk for heart disease, cancer, obesity, and innumerable other diseases by scanning his DNA for the presence of deleterious genes. For example, once a person has his DNA on disk, a physician will be able to examine the p-53 gene for mutations. Mutations of this gene cause more than half of all cancers. The normal p-53 gene signals a cell to stop dividing and thus to die. Mutation renders the gene nonfunctioning. Cells that would have died instead divide without restraint, and it is this cellular division gone awry that is cancer. Physicians may, in the future, be able to inject the normal p-53 gene into the body by attaching it to a virus, which will replicate in the body and thereby replace the faulty p-53 gene with a normal one. The risk is, however, that the body's immune system, detecting a virus, will try to kill it and by default the p-53 gene to which it is attached.

Beyond these achievements, medicine will search for nutrients that fight cancer and other diseases. For example genistein, a chemical abundant in cabbage, undermines the formation of blood cells in tumors. Starved of blood, a tumor cannot grow. The Japanese consume thirty times more genistein than do Americans, explaining why the Japanese have lower cancer rates than Americans. Surely many other nutrients that fight disease await discovery in the future.

Among its areas, research on gene therapy has focused on zebra fish, which regenerate heart muscle after it is damaged. By contrast, damage to the human heart, once it scars, is permanent. Scientists may one day identify the gene that codes for this regenerative process. This research may lead to the identification of analogous genes in us. If we do not have these genes in our genome, scientists may be able to transfer them from zebra fish to us.

In the 21st century medicine will continue to fight infectious diseases. Vaccines, if widely available, should reduce the incidence of tetanus and tuberculosis. Perhaps physicians will, using a mix of vaccines and antibiotics, eradicate tetanus and tuberculosis as they did smallpox in the 20th century. Medicine will continue its assault on AIDS, which is so insidious because HIV kills T-cells, undermining the immune system. Physicians may learn to engineer T-cells that HIV cannot latch onto and so cannot destroy. Injections of these new T-cells into the body of an AIDS patient may repair the patient's immune system so it can fight HIV rather than fall prey to it.

Microbes will fight back, though. The number of bacteria resistant to antibiotics will surely grow in the 21st century. In 1992 alone 19,000 Americans died of an infection from antibiotic-resistant bacteria. Already a strain of pneumonia is resistant to the antibiotics penicillin and cephalosporin. The active ingredient in penicillin is the beta-lactam ring, which dissolves the cell walls of bacteria. Surely researchers will find other compounds with a beta-lactam ring and thereby add another antibiotic to the arsenal of antibacterial agents. Perhaps physicians will manufacture the beta-lactam ring, inserting it into a pill. But resistant bacteria attack the beta-lactam ring by manufacturing beta-lactamase, an enzyme that breaks down beta-lactam and so renders the antibiotic ineffective. Beta-lactamase is so dangerous because plasmids manufacture it and plasmids move easily from bacterium to bacterium, transferring penicillin resistance to every bacterium that acquires a plasmid that produces beta-lactamase. Thus, physicians will need new and better antibiotics to check the spread of infectious diseases in the 21st century.

Sometime in the 21st century, researchers may discover how to increase longevity by manipulating the genes that age us. Not everyone believes such genes exist, but those who do point to the age-1 gene in nematodes. Nematodes in which the gene has been turned off live more than twice as long as those that have functioning copies of the gene. Researchers hope to find in humans the equivalent of the age-1 gene. Geron Corporation hopes to find the "immortality gene" on the assumption that a single gene controls aging.[2] The alternative is that two or more genes cause humans to age. A molecular biologist at the University of Texas Southwestern

Medical Center in Dallas has identified in humans M (mortality) 1 and M 2 genes. He hopes someday to learn how to switch these genes off. By pursuing this research, Christopher Wills, a biologist at the University of California at San Diego, believes medicine may lengthen the human life-span to 150 years.

Inevitably, medicine in the 21st century will turn up deleterious genes. Researchers intent on controlling obesity hope to find the "fat genes."[3] Other deleterious genes are fatal. Tay-Sachs, a lethal disease, derives from a single recessive gene. People who possess one gene are carriers of Tay-Sachs but manifest no symptoms. The possessor of two copies of the gene will die an excruciating death. One problem for the 21st century will be what to do with the catalog of deleterious genes. Insurance companies, if they know a person has a deleterious gene, may deny him coverage even though he is in good health. Employers may fire a worker who they discover to have a deleterious gene even though her performance on the job is satisfactory. Far from being far-fetched, this type of discrimination is already occurring. A study by Harvard University and Stanford University identified some two hundred people who had been denied insurance coverage, fired from a job, or prevented from adopting a child because they had deleterious genes.

The attempt to sort out good genes from bad will, one suspects, lead to the quest for the perfect genome. Hard, perhaps impossible, to define the perfect genome will attract Americans as the fountain of youth beckoned generations of Europeans. Whereas one may get cosmetic surgery today, one may have genetic surgery tomorrow. Scientists may be able to snip out fragments of deleterious DNA and insert functioning DNA. Parents will pay the gene manipulator who can give their children perfect DNA. Coaches will pursue athletes with the perfect DNA. The 21st century may witness the DNA divide just as the 20th century witnessed the digital divide. The DNA haves will get the good jobs and vacation at the trendiest locales. The DNA have-nots will limp along on menial jobs. Even these jobs may disappear if robots can be programmed to do them. Dating services will screen out people based on their genes, ensuring matches between only the DNA haves. A new breed of eugenicists may resurrect the claim that heredity is destiny. If Americans do not restrain this thinking, then the 21st century may witness a further widening of the gap in income and education between the haves and have-nots.

With its enormous memory, a computer makes possible the creation of a database with everyone's medical history. A physician who examines a person newly admitted to the hospital may be able to view his or her medical history, noting allergies to drugs and the like. The database would save family members as well as the patient from trying to recall these details. The database would make obsolete the lengthy pen-and-paper questionnaire that hospitals routinely give patients upon admission. Critics, however, fear that such a database would violate privacy rights and that insurance companies might gain access to it and deny a person coverage. Employers might fire workers on the basis of medical problems such a database would reveal.

Insects have long tormented humans, transmitting deadly disease to those unlucky enough to be bitten by a disease-carrying insect. Scientists may one day modify an insect so that when it mates, the female will bear only sterile offspring. Already it is possible to irradiate insects, causing them to be sterile. Sterile insects, if their population is large enough, will swamp a smaller population of fertile insects, producing either no offspring or a generation of insects incapable of having progeny. Equally important, scientists may one day modify insects so that they cannot carry and transmit diseases. This possibility may promise a future in which mosquitoes cannot transmit malaria, yellow fever, and West Nile virus and fleas cannot transmit plague. Mosquitoes hone in on us by detecting the carbon dioxide we exhale. Scientists may one day be able to engineer mosquitoes that cannot detect carbon dioxide and so will not bite us.

TOWARD OTHER PLANETS

At some time in the future we must leave earth and colonize other planets. If we remain on earth, we will die with it when the sun, swelling to a red giant, consumes the earth in fire. True, this fate will not befall humans for some five billion years, but other threats are less remote. The impact of a large meteor, such as the one that likely killed off the dinosaurs, might ignite fires worldwide and spew dirt and debris into the atmosphere, blocking out sunlight and reducing temperatures, perhaps even causing a mini ice age. Faced with these threats, we must scan the universe, immense as it is, for other planets with a climate and ecology similar to our own. It is sobering to realize that the stars in just our galaxy are so far away that we will need hundreds of years to explore by rocket just a fraction of them. We will need rocket probes that are capable of cloning themselves from the atoms of other planets and from the debris of stars—entirely fanciful at the moment. These probes, and their ever-enlarging number of clones, will populate the entire galaxy in their search for a suitable planet. Once a probe detects such a planet, we will make the journey to it by rocket, but the trip will be so long that we will not live long enough to complete it. Neither will our children. Dozens of generations of humans will pass before the rocket finally reaches land. Humans have attempted nothing like this voyage—not even Christopher Columbus's voyage compares. We will need more than science and technology to succeed. We will need courage, stamina, and luck.

Futurists, accurately or not, try to divine the future of technology. A more difficult task will be to gauge our reaction to the ever-growing use of technology in the future. Will we be happy knowing that computers are smarter than we are? Will we trust the machine-minds that can reason in ways that we cannot readily understand? Will we give up our jobs to them? Will we feel comfortable in according them quasi-human status? If we answer no to these questions, then how will we reconcile sharing the planet with these machine-minds? The human element is the unknown factor in the futurists' predictions. The Luddites smashed the machines of the Industrial Revolution. Perhaps some of us will do the same to super-computers.

Alternatively, we may embrace new technologies, particularly medical ones. The development of nanobots that patrol the bloodstream for plaque, bacteria, and viruses should thrill everyone. The growing of organs from our own stem cells has the potential to alleviate the shortage of organs for transplantation. Medical technology may increase longevity and vigor so that we grow old and robust rather than old and feeble. Life will not just lengthen, its quality will improve.

Technology can trace any of innumerable paths to the future, but we will determine which route it takes, and we will accept the consequences for what becomes of the future. The existentialists were right: we are the sum of our choices, and our choices will yield the future in all its particulars. We can take comfort from the fact that we, rather than blind chance, control our destiny.

NOTES

1. Michio Kaku, *Visions: How Science Will Revolutionize the 21st Century* (New York: Anchor Books, 1997), 143.

2. Michael G. Zey, *The Future Factor: Forces Transforming Human Destiny* New Brunswick, N.J.: Transaction, 2004), xviii.

3. Kaku, *Visions*, 229.

GLOSSARY

Aedes aegypti: A species of mosquito that carries yellow fever virus. The female transmits the virus with her bite.

Aircraft carrier: A ship with a long deck that accommodates airplanes. An aircraft carrier launches airplanes against a real or imaginary target and retrieves them at the end of their mission.

Airplane: An aircraft with wings, seats for a pilot or pilots and passengers, and either propeller-driven or jet engines.

Alternating current: A type of electricity produced by the rotation of a coil of wire in a magnetic field.

Ammonium nitrate: A solid fertilizer that releases nitrogen into the soil for absorption by plant roots.

Antibiotic: A chemical that is produced by a microbe and that is toxic to a bacterium or to a range of bacteria.

Aqua ammonia: A liquid fertilizer that releases nitrogen into the soil for absorption by plant roots.

Assembly line: A manner of organization of a factory in which the entity to be manufactured moves from worker to worker, each of whom performs a single task in contributing to the construction of the whole.

Atomic bomb: A bomb that converts a tiny amount of matter into energy. This conversion of matter into energy is instantaneous and causes a large explosion. The first atomic bombs were fission devises, but since the 1950s atomic bombs have been fusion devices.

Automobile: A means of transportation that has four wheels, a gasoline engine, and space for a driver and passengers.

Barbiturates: Derivatives of barbituric acid taken to sedate their user.

Bicycle: A frame with two wheels, pedals, a seat, handlebars, and a chain drive. Humans power a bicycle by turning the pedals.

Bubonic plague: A disease caused by the bacterium *Yersinia pestis*. Fleas transmit the bacterium through their bite. Primarily a disease of rodents, bubonic plague has several times infected humans in epidemic proportions.

Carbamates: A class of insecticide composed of nitrogen, hydrogen, carbon, and oxygen. Carbamates inactivate the enzyme acetylcholinesterase, poisoning the insect.

Cell phone: An electrical device that carries the human voice as a radio wave through space and to another phone, which receives the radio wave as sound. Unlike the telephone, the cell phone is portable.

Chlorine: A green yellow gas that was used during World War I as a poison and that is used in peacetime as a disinfectant.

Cinema: The projection onto a screen of transient images and sound.

Computer: An electronic device that carries out the instructions of software.

Corn Belt: The area bounded by Ohio in the east, Texas in the south, the Dakotas in the west, and Canada in the north. Corn is the principal crop in this region.

Corn Show: A communal event at which farmers compare their ears of corn on the basis of aesthetics.

Creationism: The belief that God created all life in its present form and in all its diversity some 6,000 years ago.

DDT: Discovered in 1939, dichlorodiphenyltrichloroethane was the first insecticide to kill insects on contact.

Direct current: A type of electricity produced by the rotation of a commutator in a magnetic field.

Double cross hybrid: The breeding of a hybrid by combining four inbreds over two generations. A farmer plants seed from the progeny of this cross.

Electrical fetal monitor: A device that monitors the heartbeat of a fetus and the onset and duration of a pregnant woman's contractions.

Electricity: The flow of electrons through a metal, through an ionic solution, or, in the case of lightning, through air.

Electric motor: A motor that uses a rotating magnetic field to turn a shaft. As the shaft turns, it imparts motion to whatever is attached to it. An electric motor can, for example, turn the blades of a fan.

Enlightenment: A philosophical movement that trumpeted the power of science and technology to improve the human condition.

Epidural: A form of anesthesia in which a physician threads a catheter into the epidural space. Drugs are then injected through the catheter to dull pain.

Eugenics: A quasi-scientific movement that attempted to reduce the complexity of human behavior and intelligence to the operation of genes and that advocated the breeding of the intelligent and industrious, with the assumption that these must be the people with good genes, and discouraged the breeding of criminals, the insane, and those with mental deficiencies, with the assumption that these must be the people with bad genes.

Evolution: The tenet that life in its current diversity descended from a single common ancestor. Over vast epochs, simpler forms diversified into a large number of complex forms of life.

Fertilizer: A soil amendment that increases crop yields.

Fluorescent light: A light that shines by generating ultraviolet light, which causes a chamber containing mercury to glow.

Freon: A hydrocarbon that is a liquid under pressure and a gas in the absence of pressure. Freon is used as a refrigerant because it cools as it expands from a liquid to a gas.

Gasoline: A distillate of petroleum. Gasoline powers automobiles, trucks, tractors, and several other types of vehicles.

Gene: A sequence of nucleotide bases that, by itself or in combination with other genes, codes for traits in plants, animals, or simpler forms of life.

Genetic engineering: The modification of an organism by inserting one or more genes into its genome or by debilitating the function of one or more genes.

Genetics: The science that seeks to clarify the effect of genes on the physical traits of organisms. In the higher animals, including humans, geneticists also attempt to study the effect of genes on mental traits.

Genome: The total number of genes in an organism. Alternatively, a genome is the total number of genes in a species.

Germ theory of disease: Asserts that pathogens, rather than bad air or arcane phenomena, cause diseases.

Great Depression: An economic contraction between 1929 and 1938 that threw Americans out of work, depressed manufacturing and agriculture, and caused some banks to close.

Haber-Bosch process: A reaction between nitrogen and hydrogen that produces ammonia, a nitrogenous fertilizer.

Hookworm: A disease caused by a worm that infects a host and inhabits the intestines, where it feeds on the food that the host eats, robbing the host of nutrition. Symptoms include lethargy, stunted growth, and mental deficiency.

Hybrid corn: Varieties of corn bred to have hybrid vigor. They were first produced by the double cross method but are now all single cross hybrids.

Hydrochlorofluorocarbons: Any of several chemicals composed of hydrogen, chlorine, fluorine, and carbon. Since the 1980s, they have been used as refrigerants. Unlike the first generation of refrigerants, hydrochlorofluorocarbons break down in the atmosphere and so do not react with the ozone layer.

Inbred corn: A type of corn that, for its genes, has been inbred to the point of being homozygous.

Incandescent light: A light that shines by heating the filament to the point that it glows.

Microwave oven: An oven that cooks food by passing microwaves through it.

Morphine: An alkaloid of opium that dulls pain and sedates the user.

Mustard gas: A compound of sulfur and chlorine that causes the skin to blister. Mustard gas was used as a poison during World War I.

Nanotechnology: An applied technology that seeks to manipulate nature at the level of one to one hundred nanometers. The goal of nanotechnology is to build everyday items atom by atom.

Narcotics: Any of a class of drug that dulls the senses, reduces pain, and induces sleep. Overuse can induce coma.

Nature study: A movement in the early 20th century to replace the monotony of textbooks in the science classroom with direct observation of nature.

Neon light: A light that shines by passing a current through a chamber containing the gas neon.

Organophosphates: A class of insecticide derived from phosphoric acid. Like carbamates, organophosphates inactivate the enzyme acetylcholinesterase, poisoning the insect.

Oxytocin: A hormone that causes the uterine muscles to contract, inducing labor in a pregnant woman.

Pellagra: A disease caused by a deficiency of niacin in the diet. Symptoms range from a mild rash to dementia.

Phosgene: A colorless gas composed of carbon, oxygen, and chlorine. Phosgene irritates the lining of the respiratory system.

Poliomyelitis: A paralytic disease caused by poliovirus.

Prontosil: The first sulfa drug. Prontosil was effective against bacterial infections.

Radio: A technology that sends and receives sound through radio waves.

Resistor: A metal that impedes the flow of electricity through it and so generates heat, which can be used in cooking.

Scientific management: A method of controlling labor by scrutinizing every facet of an industrial process to weed out inefficiency. Scientific management has as its goal the identification of the most efficient way to produce a product.

Scopolamine: A sedative derived from the roots of plants in the nightshade family.

Single cross hybrid: The breeding of a hybrid by combining two inbreds over a single generation. A farmer plants seed from the progeny of this cross.

Skateboard: A four-wheeled board that is in the shape of a small surfboard. Humans power a skateboard by pushing one foot against the ground.

Smog: A neologism made from the words *smoke* and *fog*. It comes about when airborne pollutants mix in the air to cause a haze.

Southern Corn Leaf Blight: A fungal disease of corn that in 1970 and 1971 cost U.S. farmers 15 percent of their corn crop.

Submarine: A warship that propels itself underwater.

Subway: An underground transit system.

Surfboard: A board of fiberglass used to ride waves in the ocean.

Telephone: An electrical device that carries the human voice as an electrical signal through telephone lines and to another telephone, which receives the electrical signal as sound.

Television: A modification of the cathode ray tube that projects onto a screen transient images and sound.

Tractor: A farm vehicle with four wheels, a gasoline engine, and space for the driver. Farmers use a tractor to pull agricultural implements.

Trolley: A large, electric-powered mass transit vehicle that seats dozens of people. Trolleys run above ground, whereas subways run beneath the ground.

Tuberculosis: A disease caused by the tubercle bacterium. The bacterium infects the lungs and is transmitted through coughing.

Twilight sleep: A method of rendering pregnant women semiconscious to ease the delivery of their babies.

Typhoid fever: A disease caused by the bacterium *Salmonella typhosa*. Symptoms include diarrhea, headache and inflammation of the intestines.

Vaccine: A virus or bacterium, either alive or dead, that when injected into a person provokes the body to produce antibodies against the pathogen. These antibodies confer immunity against subsequent infection by the pathogen.

Vacuum tube: A device that controls the movement of electrons through a vacuum to modify an electrical signal.

Video games: Interactive games that pit a player against images on a video screen. The violence of video games prompted the U.S. Senate twice in the 1990s to hold hearing about them.

Vitamins: Organic compounds that in minute quantities are necessary to health.

Yellow fever: A disease caused by the yellow fever virus. Initial symptoms mimic those of the flu. Yellow fever attacks the kidneys and liver and, in fatal cases, causes internal bleeding.

Yellow fever virus: A flavivirus that originated in Africa and is transmitted by several species of mosquito, the best known being *Aedes aegypti*.

FURTHER READING

Allen, Garland E. "Eugenics and American Social History." *Genome* 31 (1989): 885–889.

———. "Genetics, Eugenics and Class Struggle." *Genetics* 79 (1975): 29–45.

———. "The Misuse of Biological Hierarchies: The American Eugenics Movement, 1900–1940." *History and Philosophy of the Life Sciences* 5 (1984): 105–128.

Amyes, Sebastian G. B. *Magic Bullets, Lost Horizons: The Rise and Fall of Antibiotics.* New York: Taylor and Francis, 2001.

Apple, Rima D. *Vitamania: Vitamins in American Culture.* New Brunswick, N.J.: Rutgers University Press, 1996.

Apple, Rima D., and Janet Golden, eds. *Mothers and Motherhood: Readings in American History.* Columbus: Ohio State University Press, 1997.

Atkin, J. Myron, and Paul Black. *Inside Science Education Reform: A History of Curricular and Policy Change.* New York: Teachers College Press, 2003.

Ballantine, Richard. *Ultimate Bicycle Book.* New York: Dorling Kindersley, 1998.

Bell, Brian. *Farm Machinery.* Tonbridge, UK: Farming Press Books, 1999.

Bernardini, Carlo, Carlo Tarsitani, and Matilde Vincentini, ed. *Thinking Physics for Teaching.* New York: Plenum Press, 1995.

Bilstein, Roger. *Flight in America, 1900–1983: From the Wrights to the Astronauts.* Baltimore: Johns Hopkins University Press, 1984.

Bordley, James, and A. McGehee Harvey. *Two Centuries of American Medicine, 1776–1976.* Philadelphia: Saunders, 1976.

Brandt, Allan M. *No Magic Bullet: A Social History of Venereal Disease in the United States Since 1880.* New York: Oxford University Press, 1985.

Brinkley, Douglas. *Wheels for the World: Henry Ford, His Company, and a Century of Progress, 1903–2003.* New York: Viking, 2003.

Brisick, Jamie. *Have Board Will Travel: The Definitive History of Surf, Skate, and Snow.* New York: HarperEntertainment, 2004.

Budiansky, Stephen. *Air Power: The Men, Machines, and Ideas That Revolutionized War, from Kitty Hawk to Gulf War II.* New York: Viking, 2004.

Carrigan, Jo Ann. *The Saffron Scourge: A History of Yellow Fever in Louisiana, 1796–1905.* Lafayette: University of Southwest Louisiana, 1999.

Cavin, Ruth. *Trolleys: Riding and Remembering the Electric Interurban Railways.* New York: Hawthorn Books, 1976.

Clark, Paul. *The Phone: An Appreciation.* San Diego, Calif.: Laurel Glen, 1997.

Cochrane, Willard W. *The Development of American Agriculture: A Historical Analysis.* Minneapolis: University of Minnesota Press, 1993.

Cooney, Brian. *Posthumanity: Thinking Philosophically About the Future.* Lanham, MD: Rowman and Littlefield, 2004.

Cooper, Kenneth H. *Aerobics.* New York: Bantam Books, 1968.

Cooper, Mildred, and Kenneth H. Cooper. *Aerobics for Women.* Philadelphia: Lippincott, 1972.

Cornish. Edward, ed. *Exploring Your Future: Living, Learning and Working in the Information Age.* Bethesda, Md.: World Future Society, 1996.

Cuban, Larry, "The Integration of Modern Sciences into the American Secondary School, 1890s–1990s." *Studies in Philosophy and Education* 18 (1999): 67–87.

Cudahy, Brian J. *Under the Sidewalks of New York: The Story of the Greatest Subway System in the World.* New York: Viking Penguin, 1989.

Davis, James. *Skateboarding Is Not a Crime: 50 Years of Street Culture.* Richmond Hills, ON: Firefly Books, 2004.

DeBoer, George E. *A History of Ideas in Science Education: Implications for Practice.* New York: Teachers College Press, 1991.

Dickerson, James. *Yellow Fever: A Deadly Disease Poised to Kill Again.* Amherst, N.Y.: Prometheus Books, 2006.

Diehl, Lorraine B. *Subways: The Tracks That Built New York City.* New York: Clarkson Potter, 2004.

Donahue, David M. "Serving Students, Science, or Society? The Secondary School Physics Curriculum in the United States, 1930–65." *History of Education Quarterly* 33 (1993): 321–352.

Douglas, Susan J. *Listening In: Radio and the American Imagination.* Minneapolis: University of Minnesota Press, 2004.

Dunbar, Charles S. *Buses, Trolleys and Trams.* London: Hamlyn, 1967.

Ellis, John. *The Social History of the Machine Gun.* Baltimore: Johns Hopkins University Press, 1986.

Etheridge, Elizabeth W. *The Butterfly Caste: A Social History of Pellagra in the South.* Westport, Conn.: Greenwood Press, 1972.

Fussell, Betty. *The Story of Corn.* New York: Knopf, 1992.

Gandy, Matthew. *Concrete and Clay: Reworking Nature in New York City.* Cambridge, Mass.: MIT Press, 2002.

Gilman, Charlotte Perkins. *The Home, Its Work and Influence.* New York: McClure Phillips, 1903.

Goer, Henci. *The Thinking Woman's Guide to a Better Birth.* New York: Berkley Publishing Group, 1999.

Goertzel, Ben. *The Path to Posthumanity: 21st Century Technology and Its Radical Implications for Mind, Society and Reality.* Bethesda, Md.: Academica Press, 2006.

Graham, Stephen, and Simon Marvin. *Splintering Urbanism: Networked Infrastructures, Technological Mobilities and the Urban Condition.* London: Routledge, 2001.

Halter, Ed. *From Sun Tzu to Xbox: War and Video Games.* New York: Thunder's Mouth Press, 2006.

Heller, Charles E. *Chemical Warfare in World War I: The American Experience, 1917–1918.* Fort Leavenworth, Kans.: Combat Studies Institute, 1985.

Herz, J. C. *Joystick Nation: How Videogames Ate Our Quarters, Won Our Hearts, and Rewired Our Minds.* Boston: Little, Brown, 1997.

Howe, Eric Michael. "Untangling Sickle-Cell Anemia and the Teaching of Heterozygote Protection." *Science and Education* (2005): 1–19.

Howell, Joel D. *Technology in the Hospital: Transforming Patient Care in the Early Twentieth Century.* Baltimore: Johns Hopkins University Press, 1995.

Hudson, Kenneth and Julian Pettifer, *Diamonds in the Sky: A Social History of Air Travel.* London: British Broadcasting Corporation, 1979.

Hughes, Thomas P. *Networks of Power: Electrification in Western Society, 1880–1930.* Baltimore: Johns Hopkins University Press, 1983.

Hughes, Tim, Nick Hamlyn, and Robert Barbutt. *The Bike Book.* New York: Gallery Books, 1990.

Jarrett, Philip. *Ultimate Aircraft.* New York: Dorling Kindersley, 2000.

Kaku, Michio. *Visions: How Science Will Revolutionize the 21st Century.* New York: Anchor Books, 1997.

Kaledin, Eugenia. *Daily Life in the United States, 1940–1959: Shifting Worlds.* Westport, Conn.: Greenwood Press, 2000.

King, Kenneth P. *Technology, Science Teaching and Literacy: A Century of Growth.* New York: Kluwer Academic, 2001.

King, Lucien, ed. *Game On: The History and Culture of Videogames.* New York: St. Martin's Press, 2002.

Kyvig, David E. *Daily Life in the United States, 1920–1939: Decades of Promise and Pain.* Westport, Conn.: Greenwood Press, 2001.

Laguerre, Michel S. *The Digital City: The American Metropolis and Information Technology.* Macmillan, 2005.

Levinson, Paul. *Cellphone: The Story of the World's Most Mobile Medium and How It Has Transformed Everything!* New York: Palgrave Macmillan, 2004.

Lewis, Rosalind. *Information Technology in the Home: Barriers, Opportunities and Research Directions.* Santa Monica, Calif.: RAND, 2000.

Lloyd, Les, ed. *Teaching with Technology: Rethinking Tradition.* Medford, N.J.: Information Today, 2000.

Martin, James. *The Meaning of the 21st Century: A Vital Blueprint for Ensuring Our Future.* New York: Riverhead Books, 2006.

Martin, Michael. *History of Skateboarding: From the Backyard to the Big Time.* Mankato, Minn.: Capstone High-Interest Books, 2002.

Marty, Myron A. *Daily Life in the United States, 1960–1990: Decades of Discord.* Westport, Conn.: Greenwood Press, 1997.

Mattern, Joanne. *The History of Radio.* Berkeley Heights, N.J.: Enslow, 2002.

McShane, Clay. *Down the Asphalt Path: American Cities and the Coming of the Automobile.* New York: Columbia University Press, 1994.

Melosi, Marvin V. *The Sanitary City: Urban Infrastructure in America from Colonial Times to the Press.* Baltimore: Johns Hopkins University Press, 2000.

Mercer, David. *The Telephone: The Life Story of a Technology.* Westport, Conn.: Greenwood Press, 2006.

Miller, John Anderson. *Fares Please! A Popular History of Trolleys, Horse-Cars, Street-Cars, Buses, Elevateds, and Subways.* New York: Dover, 1960.

Mills, Robert K., ed. *Implement and Tractor: Reflections on 100 Years of Farm Equipment*. Marceline, Missouri: Intertec, 1986.

Montgomery, Scott L. *Minds for the Making: The Role of Science in American Education, 1750–1990*. New York: Guilford Press, 1994.

Nelkin, Dorothy. *The Creation Controversy: Science or Scripture in the Schools*. Boston: Beacon Press, 1984.

Nye, David E. *Electrifying America: Social Meanings of the New Technology, 1880 1940*. Cambridge, Mass.: MIT Press, 1990.

Oliver, Smith Hempstone. *Wheels and Wheeling: The Smithsonian Cycle Collection*. Washington, D.C.: Smithsonian Institution Press, 1974.

Pauly, Philip J. "The Development of High School Biology: New York City, 1900–1925." *Isis* 82 (1991): 662–688.

Perkins, John H. *Insects, Experts, and the Insecticide Crisis: The Quest for New Pest Management Strategies*. New York: Plenum Press, 1982.

Peterson, Peter A., and Angelo Bianchi, eds. *Maize Genetics and Breeding in the 20th Century*. River Edge, N.J.: World Scientific, 1999.

Platt, Harold L. *Shock Cities: The Environmental Transformation and Reform of Manchester and Chicago*. Chicago: University of Chicago Press, 2005.

Poole, Steven. *Trigger Happy: Videogames and the Entertainment Revolution*: New York: Arcade Publishers, 2000.

Resnick, Rosalind, and Heidi Anderson. *A Pocket Tour of Shopping on the Internet*. San Francisco: SYBEX, 1996.

Rothman, Barbara Katz. *Recreating Motherhood: Ideology and Technology in a Patriarchal Society*. New York: Norton, 1989.

Rudge, David Wyss. "Does Being Wrong Make Kettlewell Wrong for Science Teaching?" *Journal of Biological Education* 35 (2000): 5–11.

Rudolph, John L. "Turning Science Education to Account: Chicago and the General Science Movement in Secondary Education, 1905–1920." *Isis* 96 (2005): 353–389.

Rudolph, John L. "Epistemology for the Masses: The Origins of the "Scientific Method" in American Schools." *History of Education Quarterly* 45 (2005): 341–375.

Rudolph, John L. *Scientists in the Classroom: The Cold War Reconstruction of American Science Education*. New York: Palgrave Macmillan, 2002.

Rybczynski, Witold. *City Life*. New York: Simon & Schuster, 1995.

———. *Home: A Short History of an Idea*. New York: Viking, 1986.

Sandelowski, Margarete. *Pain, Pleasure and American Childbirth: From the Twilight Sleep to the Read Method, 1914–1960*. Westport, Conn.: Greenwood Press, 1984.

Shneiderman, Ben. *Leonardo's Laptop: Human Needs and the New Computing Technologies*. Cambridge, Mass.: MIT Press, 2002.

Sidwells, Chris. *Complete Bike Book*. New York: Dorling Kindersley, 2005.

Silberglitt, Richard, Philip S. Anton, David R. Howell, and Anny Wong. *The Global Technology Revolution 2020, In-Depth Analyses: Bio/Nano/Materials/Information Trends, Drives, Barriers, and Social Implications*. Santa Monica, Calif.: RAND, 2006.

Smith, Anthony. *Machine Gun: The Story of the Men and the Weapon That Changed the Face of War*. New York: St. Martin's Press, 2003.

Spector, Ronald H. *At War, At Sea: Sailors and Naval Warfare in the Twentieth Century.* New York: Viking, 2001.

Spock, Benjamin. *Baby and Child Care.* New York: Hawthorn Books, 1968.

Sprague, George F., and John W. Dudley, eds. *Corn and Corn Improvement.* Madison, Wis.: American Society of Agronomy, 1988.

Starr, Paul. *The Social Transformation of American Medicine.* New York: Basic Books, 1982.

Straubhaar, Joseph, and Robert LaRose. *Communications Media in the Information Society.* Belmont, Calif.: Wadsorth, 1996.

Strode, George K., ed. *Yellow Fever.* New York: McGraw-Hill, 1951.

Tarr, Joel A., ed. *The City and Technology.* Beverly Hills, Calif.: Sage, 1979.

Terry, Jennifer, and Melodie Calvert, eds. *Processed Lives: Gender and Technology in Everyday Life.* New York: Routledge, 1997.

Terzian, Sevan. "Science World, High School Girls, and the Prospect of Science Careers." *History of Education Quarterly* 46 (2006): 73–99.

Time Books. *Great Inventions: Geniuses and Gizmos; Innovation in Our Time.* New York: Time Books, 2003.

Tolley, Kim. *The Science Education of American Girls: A Historical Perspective.* New York: RoutledgeFalmer, 2003.

Trask, Benjamin. *Fearful Ravages: Yellow Fever in New Orleans, 1796–1905.* Lafayette: University of Louisiana at Lafayette, 2005.

Warner, John Harley, and Janet A. Tighe, eds. *Major Problems in the History of American Medicine and Public Health: Documents and Essays.* Boston: Houghton Mifflin, 2001.

Webb, George E. *The Evolution Controversy in America.* Lexington: University Press of Kentucky, 1994.

Weyland, Jocko. *The Answer Is Never: A Skateboarder's History of the World.* New York: Grove Press, 2002.

Whyte, William H. *City: Rediscovering the Center.* New York: Doubleday, 1988.

Wieners, Brad and David Pescovitz. *Reality Check.* San Francisco: Hardwired, 1996.

Winkowski, Fred. *100 Planes, 100 Years: The First Century of Aviation.* New York: Smithmark, 1998.

Yager, Robert, ed. *Science/Technology/Society as Reform in Science Education.* Albany: State University of New York Press, 1996.

Zey, Michael G. *The Future Factor: Forces Transforming Human Destiny.* New Brunswick, N.J.: Transaction, 2004.

INDEX

About the Author

CHRISTOPHER CUMO is a freelance scholar. His specialty is American regional history. He has published two books on midwestern history: *Seeds of Changes,* and *A History of the Ohio Agricultural Experiment Station, 1882-1997.* He has a Ph.D in history from the University of Akron.